OXFORD LOGIC GUIDES: 31

General Editors

DOV GABBAY
ANGUS MACINTYRE
DANA SCOTT

OXFORD LOGIC GUIDES

---

1. Jane Bridge: *Beginning model theory: the completeness theorem and some consequences*
2. Michael Dummett: *Elements of intuitionism*
3. A. S. Troelstra: *Choice sequences: a chapter of intuitionistic mathematics*
4. J. L. Bell: *Boolean-valued models and independence proofs in set theory* (1st edition)
5. Krister Seberberg: *Classical propositional operators: an exercise in the foundation of logic*
6. G. C. Smith: *The Boole–De Morgan correspondence 1842–1864*
7. Alec Fisher: *Formal number theory and computability: a work book*
8. Anand Pillay: *An introduction to stability theory*
9. H. E. Rose: *Subrecursion: functions and hierarchies*
10. Michael Hallett: *Cantorian set theory and limitation of size*
11. R. Mansfield and G. Weitkamp: *Recursive aspects of descriptive set theory*
12. J. L. Bell: *Boolean-valued models and independence proofs in set theory* (2nd edition)
13. Melvin Fitting: *Computability theory: semantics and logic programming*
14. J. L. Bell: *Toposes and local set theories: an introduction*
15. R. Kaye: *Models of Peano arithmetic*
16. J. Chapman and F. Rowbottom: *Relative category theory and geometric morphisms: a logical approach*
17. Stewart Shapiro: *Foundations without foundationalism*
18. John P. Cleave: *A study of logics*
19. R. M. Smullyan: *Gödel's incompleteness theorem*
20. T. E. Forster: *Set theory with a universal set: exploring an untyped universe*
21. C. McLarty: *Elementary categories, elementary toposes*
22. R. M. Smullyan: *Recursion theory for metamathematics*
23. Peter Clote and Jan Krajíček: *Arithmetic, proof theory, and computational complexity*
24. A. Tarski: *Introduction to logic and to the methodology of deductive sciences*
25. G. Malinowski: *Many valued logics*
26. Alexandre Borovik and Ali Nesin: *Groups of finite Morley rank*
27. R. M. Smullyan: *Diagonalization and self-reference*
28. Dov M. Gabbay, Ian Hodkinson, and Mark Reynolds: *Temporal logic: Mathematical foundations and computational aspects (Volume I)*
29. Saharon Shelah: *Cardinal arithmetic*
30. Erik Sandewall: *Features and fluents: Volume I: A systematic approach to the representation of knowledge about dynamical systems*
31. T. E. Forster: *Set theory with a universal set: exploring an untyped universe* (2nd edition)

Set Theory with a Universal Set

Exploring an Untyped Universe

Second Edition

T. E. FORSTER

Department of Pure Mathematics and Mathematical Statistics, University of Cambridge

CLARENDON PRESS · OXFORD

1995

Oxford University Press, Walton Street, Oxford OX2 6DP
Oxford New York
Athens Auckland Bangkok Bombay
Calcutta Cape Town Dar es Salaam Delhi
Florence Hong Kong Istanbul Karachi
Kuala Lumpur Madras Madrid Melbourne
Mexico City Nairobi Paris Singapore
Taipei Tokyo Toronto
and associated companies in
Berlin Ibadan

Oxford is a trade mark of Oxford University Press

Published in the United States by
Oxford University Press Inc., New York

First published 1992
Second edition 1995

A catalogue record for this book is available from the British Library

Library of Congress Cataloging in Publication Data
(Data applied for)

ISBN 0 19 851477 8

Typeset by the author using LaTeX

Printed in Great Britain by
Biddles Ltd, Guildford and King's Lynn

PREFACE TO THE FIRST EDITION

This book is an *essay*: not a monograph, or a textbook, but an essay. It is intended to be a good read for those people who are already interested in this topic (or think they might become interested in it) rather than a comprehensive treatment for people who wish to master it, and a reference work for those who already have. Unfortunately there simply are not enough such people! It was originally going to be such a textbook, but it gradually became clear that the logistical problems of the collaboration that this entailed were going to be insuperable, and it has taken its present form of an essay biased in the direction of my interests. Regrettable though this bias is, it is unavoidable given that it is no longer possible for any one person to write a comprehensive book on this subject.

It is never a good time to write a book about anything, for there is always somewhere a rapidly moving target that one has to take one's eye off in order to put pen to paper. In the present case one such rapidly moving target is Richard Kaye, who has recently become interested in the consistency of subsystems of *NF* (Quine's *New Foundations*), and who has a body of unpublished work that cannot conveniently be treated here. I have attempted to touch all the major areas of set theory with a universal set, though I admit to not having made any serious attempt with Skala, and I have kept a low profile where positive set theory is mentioned. My inclination to give this last item an extensive treatment—never very strong—was dealt a mortal blow by seeing Weydert's Ph.D. thesis and finally extinguished by sight of the proofs of the article by Forti and Hinnion [1989] which between them contain as good an introduction to this subject as I could hope to write, or the reader could reasonably wish to see. There is no point in attempting to duplicate good expositions. Positive set theory, too, is a rapidly moving target, and the best I can do is provide a brief introduction and direct readers on their way through the bibliography. The result is a book that concentrates heavily on *NF* and is much more like a second edition of Forster [1983a] than I would have wished. To a certain extent, this is unavoidable: *NF* is a much richer and more mysterious system than the other set theories with a universal set, and there are large areas in its study (e.g. the reduction of the consistency question) which have no counterparts elsewhere in the study of set theories with $V \in V$.[1] There just is a great deal more to say about *NF* than about the other systems. Although I

[1] This is just technical slang for "V, the universe, is a set".

have not attempted to discuss all open questions, I have tried to cover those that seemed interesting to me, to explain why they seemed interesting, and to try to give pointers for people who wish to pursue topics that I do not. The coverage of Rieger–Bernays permutation models is the most comprehensive to be had, and the treatment of the theories in the tradition of Church's universal set theory is likely to remain the most extensive of any in print until Sheridan finishes [199?] and rewrites it for publication. Much that is complicated is not proved here—for example, at one point, readers are assumed to be familiar with the Ehrenfeucht–Mostowski theorem and at another with the Keisler–Shelah ultrapower lemma—and yet some quite elementary material is treated in considerable detail. This is not perverse: after my prejudices have had their say, what determines the amount of coverage a topic gets is not its intrinsic difficulty but the extent to which it is both unfamiliar and not well covered in any highly visible text. The result is probably a rather odd document: much of what is within is intelligible even to people who have so little logic that much of the motivation of the text will not be apparent to them. Readers who are not specialist set theorists should bear in mind that some parts (at least) of this volume should be read in conjunction with a good introductory graduate text on axiomatic set theory, of which there are many.

A brief, partly historical, survey is provided in section 1.3 for those who want a free sample before buying the product.

I would like to thank the Science and Engineering Research Council and the University of Cambridge for employing me while this was being written. I would also like to thank members of the Cambridge and Wellington Logic Seminars, and especially the Séminaire NF, for helpful comments and for their stimulating companionship over the years. Finally I owe a special debt of gratitude to Richard Kaye, not only for permission to use our joint work but also for reading earlier drafts of this text (at considerable cost in his time) and compelling me to clarify points about which I was being slovenly. I am similarly indebted to Maurice Boffa, who read through the penultimate drafts for OUP at similar cost to his time. If this book succeeds in its purpose, it is in a large part due to them.

Cambridge T.E.F.
March 1991

PREFACE TO THE SECOND EDITION

No effort has been made to rewrite the book to explain fully all recent research in set theory with a universal set, though all such research known to me is at least minuted. I have resisted the temptation to slant the material towards my current research interest: constructive *NF*. Minor errors have been corrected. The only major error to have surfaced is the claim that $(\forall n \in \mathbb{N})(n \leq Tn) \longleftrightarrow \Diamond\exists V_\omega$. The right hand side should have been "$\Diamond\{V_n : n \in \mathbb{N}\}$ exists". The major unforced change is the complete revision of the chapter on the set theories of Church and Mitchell, though the section on $\in$-games has been revised and enlarged.

Various minor *aperçus* have been inserted, and I am grateful to Robert Jones of Wegberg for finding errors and omissions in the bibliography. I am extremely grateful to members of the *NF* net for the time they spent reading through late draughts of the second edition to check for errors and infelicities: this book owes not only a lot of its material to them (and this is acknowledged in line) but also its present tolerable presentation. It is a pleasure to have this opportunity to thank them here for their help and companionship. Richard Kaye is owed a particular debt of gratitude for checking last minute changes to the section on $\in$-games.

Cambridge T.E.F.
March 1995

CONTENTS

1	**Introduction**	1
1.1	Annotated definitions	4
	1.1.1 Quantifier hierarchies	5
	1.1.2 Mainly concerning type theory	6
	1.1.3 Other definitions	9
	1.1.4 Theories	10
1.2	Some motivations and axioms	11
	1.2.1 Sets as predicates-in-extension	11
	1.2.2 Sets as natural kinds	21
1.3	A brief survey	22
1.4	How do theories with $V \in V$ avoid the paradoxes?	24
1.5	Chronology	25
2	***NF* and related systems**	26
2.1	*NF*	26
	2.1.1 The axiom of counting	30
	2.1.2 Boffa's lemma on n-formulae, and the automorphism lemma for set abstracts	33
	2.1.3 Miscellaneous combinatorics	35
	2.1.4 Well-founded sets	40
2.2	Cardinal and ordinal arithmetic	44
	2.2.1 Some remarks on inductive definitions	55
	2.2.2 Closure properties of small sets	57
2.3	The Kaye–Specker equiconsistency lemma	58
	2.3.1 NF_3	65
	2.3.2 *NFU*	67
	2.3.3 Lake's model	72
	2.3.4 KF	72
2.4	Subsystems, term models, and prefix classes	83
2.5	The converse consistency problem	89
3	**Permutation models**	92
3.1	Permutations in *NF*	96
	3.1.1 Inner permutations in *NF*	97
	3.1.2 Outer automorphisms in *NF*	119
3.2	Applications to other theories	121

4 Church–Oswald models 122

4.1 Oswald's model 122

4.2 Low sets 124

4.2.1 Other definitions of low 125

4.3 $\mathcal{P}$-extensions and permutation models 126

4.3.1 $\mathcal{P}$-extensions 126

4.3.2 Hereditarily low sets and permutation models 127

4.3.3 Permutation models of CO structures 129

4.4 Two applications 130

4.4.1 An elementary example 130

4.4.2 $\mathcal{P}$-extending models of Zermelo to models of *NFO* 132

4.5 Church's model 136

4.6 Mitchell's set theory 139

4.7 Conclusions 140

5 Open problems 143

5.1 Permutation models and quantifier hierarchies 143

5.2 Cardinals and ordinals in *NF* 144

5.3 KF 144

5.4 Other subsystems 145

5.5 Well-founded extensional relations 145

5.6 Term models 146

5.7 Miscellaneous 146

Bibliography 148

Index of definitions 161

Author index 163

General index 164

1

INTRODUCTION

It is often assumed that Russell's paradox shows that the universe cannot be a set, but this follows only if every subclass of a set is a set. So, if we want the universe to be a set (hereafter "$V \in V$"), we must somehow forestall the application of comprehension-like schemes to big[2] sets, lest we enable the diagonal construction to go through. This does not exclude the possibility that, for some suitable notion of "small", the small sets in such a system form a model of *ZF*. In some of the systems described here this possibility is made explicit in an axiom whose relative consistency can be proved. Thus the extra generality of $V \in V$ need not cost anything.

There are natural reasons for considering set theories with $V \in V$. Perhaps the oldest and best is the point of view that set membership can be seen as an allegory for predication, so that sets are deemed to arise as extensions of predicates. Set membership is to be seen as a cleaned-up formal version of predication, and our study of the natural object can benefit from attention to the formal model in the usual way. Perhaps a better way to put this is: sets are predicates-in-extension. (It is worth noting parenthetically that this opens up a very deep difference between sets and multisets. There does not seem to be an intensional x such that multisets are x-in-extension.) On this viewpoint, which tends to be held by logicists (even though it is clearly not part of the official logicist picture nor is this view held only by logicists), one would naturally want the universe to be a set, since, if one thing is certain in this life, it is that everything is identical with itself, so we cannot avoid having the universe as the extension of the predicate of self-identity. Thus set theories with a universal set are likely to be of interest to anyone who wants to see what a genuinely untyped universe must look like. (The fact that *NF* has its roots in Russellian type theory only adds piquancy to this development.) Another topical reason for studying these theories is the hope that they will furnish natural and illuminating models both of untyped and of polymorphic λ-calculus. Indeed, for a number of readers, the principal attraction of a volume such as this is

[2] Cardinals such as $\overline{\overline{V}}$ or sets that are the extension of a class abstract that would have to be a proper class in *ZF* (Zermelo–Fränkel set theory) or *GB* (Gödel–Bernays set theory), which are clearly enormous in this straightforward but novel sense, we shall call "big" cardinals (or sets) rather than "large" cardinals (or sets) to avoid confusion with *ZF* usage. 'Big' and 'large' are now both technical terms!

that a good understanding of Russellian type theory—which is after all a special case of the type theory in Church [1940] (which confusingly is also called "simple type theory"!)—and its one-sorted relatives may provide a useful rehearsal for investigations into various λ-calculi.[3] Yet another recent point of view (associated with Peter Aczel) is that the $\in$-diagram of a set can be seen as coding the course of a computation. (This leads us too far, since it also involves ill-founded set theories *without* a universal set and for them there is no room at the inn.) Those interested may consult Aczel [1988]. Finally we may note that recent developments in the study of the consistency problem for *NF* have thrown up some novel theories of well-founded sets that are incompatible with *ZF*, so the study of these supposedly outlandish theories may yet have something to tell us about well-founded sets.

For a long time set theory with a universal set has been the Cinderella of logic. It has been ignored—as late as 1988 the *Journal of Symbolic Logic* could devote an entire issue to the foundations of mathematics without even mentioning it—and misrepresented. In his introduction to Aczel [1988] Barwise wrote: "Just as there used to be complaints about referring to complex numbers as numbers, so there are objections to referring to non-well-founded sets as sets ...". The implication of this comparison is that ill-founded sets are a titillating and fertile novelty (like complex numbers) to which we should extend a warm welcome in the same way. Polemically of course Barwise is right; historically he is quite wrong: set theory was *born* ill-founded (had it not been, Russell's and Cantor's paradoxes would not have been discovered when they were, or indeed, at all!) and well-founded set theory is merely a pampered part of it. It is a bit like someone familiar only with the additive groups of reals and rationals deciding that non-abelian groups are really jolly interesting. This is of course the kind of mistake one would expect to be made by people who are brought up in the *ZF* world, since in that culture the paradoxes are viewed as large holes in the ground that one might fall into, so that, once one has found even one way of avoiding them, one immediately changes the subject. However, it is *always* a mistake to think of *anything* in mathematics as a *mere* pathology, for there are no such things in mathematics. The view behind this book is that one should think of the paradoxes as supernatural creatures, oracles, minor demons, etc. (or perhaps the *Aleph* in the eponymous story by Borges)—on whom one should keep a weather eye in case they make prophecies or by some other means inadvertently divulge information from another world not normally obtainable otherwise. One should approach

[3] These readers should be warned that Church's set theory with a universal set which is discussed here has no special relationship with his simple type theory.

them as closely as is safe, and from as many different angles as possible.[4] Hence set theory with a universal set (which is another angle), and hence this book.

This disregard by the mathematician-in-the-street has persisted despite the fact that two of the century's most distinguished logicians—Quine and Church—considered the topic sufficiently worthy of attention for them to make serious contributions to it. Indeed, research has now been proceeding in this area for some 50 years (the 50th anniversary of Quine's [1937a] was commemorated by a week-long meeting at Oberwolfach which practically all current workers in *NF* attended) and the literature now runs to nearly a hundred and fifty items, over a hundred of them since 1970. The large majority of this work concerns *NF* and its subsystems, and this is reflected in the coverage here. The point is not that *NF* is the only interesting system, but rather that it is the oldest (the Church and Mitchell systems appeared in 1974 and 1976) and it is the most *idiomatic* set theory with a universal set. Through the poverty of the manipulations of *big* sets that they allow, the systems of Church and Mitchell betray their origins in consistency proofs arising from manipulations of models of *ZF*: for example, although both systems allow that all complements of well-founded sets exist, neither allows us to form the cartesian product of two such sets. In their model theory, the Church and Mitchell systems resemble each other very much more than either resembles the *NF* family. There are very few techniques that can be applied profitably to both families.

Positive set theory is a more recent development still. Actually "positive" is not a good description: Weydert writes of "topological set theory". The idea is that every class is approximated by a set and that the approximation operator is idempotent. The idea of sets approximating classes appears first in Skala [1974a,b]. The idea that this offers an escape from the paradoxes is beautifully encapsulated in the last line of the "argument" sketch of Monty Python. In it, the official who is paid to argue with a client, and chooses to argue about whether the client has paid or not, replies to the challenge "If I haven't paid, why are you still arguing?" with the words "I might be arguing in my spare time!" The set most closely approaching the Russell class might be a superset of it.

Weydert takes this further. Sets are thought of as classes that are closed in some topology (closure is idempotent, after all). This has a rich model theory, but the connection between the constructions and the resulting comprehension axioms are not clear. "Positive" formulae (which give rise to comprehension axioms) seem, to me at least, to be hard to recognize.

[4] I admit to a lingering sympathy for those on the lunatic fringe of our subject who are obsessed with the paradoxes. For them the paradoxes are like the *zahir* of Borges' eponymous story behind which—as he suggests—we might find God.

Set theories with $V \in V$ retain some ancestral features lost in the more specialized theories of well-founded sets. This becomes apparent when we consider the possibility of standard versions of strong axioms. A strong axiom is one that says that V is closed under some operation, or that something highly improbable happens. Either way it boils down to saying that the universe is large, and since *ZF* nails it down at one end, the other is a long way away. Thus in well-founded theories like *ZF*, whatever we started off trying to say, the result is always liable to be (equivalent to) a *large cardinal axiom.* The fact that interesting allegations in *ZF* tend to be related to large cardinal axioms is thus revealed to be merely an artefact of the axiom of foundation, which determines that large cardinals are, so to speak, the local currency in which information is traded and strength is denominated. The situation in set theories with $V \in V$ is not so straightforward.

1.1 Annotated definitions

We should take note early on that we shall not be considering in detail any theories with proper classes. This is for much the same reasons that such theories are nowadays largely disregarded by students of *ZF*. Quine's *ML* ("Mathematical Logic"), which stands in the same relation to *NF* that *GB* does to *ZF*, will be mentioned *en passant* in section 2.2, but only as a terrible warning.

Frequently it will be necessary to distinguish between formulae and what they are used to say; single quotation marks will be used for this. At times, Quine quotes—$\ulcorner\Phi\urcorner$—will be used as well. In this case the 'Φ' is actually a metavariable taking formulae as values (see Quine [1951a]).

In *NF* at least, it has become customary (following Rosser [1953a]) to use the upper case to denote extensions of predicates. Thus '*NC*' for the set of cardinal numbers, '*NO*' for the set of ordinal numbers, and '*NCI*' for the set of cardinal numbers of infinite sets. 'Λ' (an upside-down 'V') denotes the empty set, '0' denotes the cardinal number 0 (which may not be the same thing!), and '1-1' denotes the set of 1-1 functions. '*On*' will continue to denote the class of all von Neumann ordinals (typically in *ZF*). V_α (some writers prefer 'R_α') is the αth level of the cumulative hierarchy of von Neumann. A finite set is one whose cardinal is in $\mathbb{N}$, the natural numbers.

Following Russell, Quine, and Gödel, we will write '$F(x)$' for a sentence asserting that x has F, ' $f`x$ ' for the (single) value that the function f gives to the argument x, and ' $f``x$ ' for the set of values of f for arguments in x. Failure to respect these distinctions would make definitions like that of j below quite unintelligible. Sometimes '(' and ')' will appear, but only in the usual way to distinguish $(f`g)`x$ from $f`(g`x)$.

The ordered pair of x and y will be *notated* "$\langle x, y\rangle$". It will be a matter of discussion how $\langle x, y\rangle$ should be *implemented.* (x, y) is the transposition of x and y, and $(a_1, a_2 \ldots a_n)$ is the cycle sending a_1 to a_2 ... and a_n to a_1. Composition of two functions f and g is $f \circ g$ (do g first and then f). The '$\circ$' will be omitted where no confusion results.

$TC(x)$ is the transitive closure of x. This is not always defined in the set theories we consider.

$card(\alpha)$, when α is an ordinal, is the cardinal number associated with α, the size of any set possessing a well-ordering of order-type α. *Alephs* are cardinals of well-ordered sets.

The sequence of *beth* numbers is defined by recursion: $\beth_0 = \aleph_0$, $\beth_{\alpha+1} = 2^{\beth_\alpha}$, taking limits at limit ordinals. Also beth numbers are defined with arguments as well as subscripts: $\beth_0`\beta = \beta$, $\beth_{\alpha+1}`\beta = 2^{(\beth_\alpha`\beta)}$, taking limits at limit ordinals. Thus, if there is no argument to a beth number, it is assumed to be $\aleph_0$.

Of course, how we implement cardinal arithmetic in any of these theories is a matter to be decided later.

Where we talk about games, "Wins" with capital 'W' means "has a winning strategy for". '■' indicates the end of a proof.

1.1.1 *Quantifier hierarchies*

A sentence in prenex normal form is said to be $\forall^n$ if its prefix contains precisely n universal quantifiers. $\forall^*$ is the prenex class containing all formulae whose prefix contains nothing but universal quantifiers. Similarly for existential quantifiers. The class $\forall^*\exists^*$ is analogously the class of well-formed formulae (wffs) whose prefix consists of universal quantifiers followed by existential. We will sometimes use subscripts to count *blocks* of quantifiers, and accordingly write "$\forall_2$" instead of "$\forall^*\exists^*$" etc.

The hierarchy of formulae in Levy [1965] will also be of interest to us here. The class of Δ_0^{Levy} formulae is the least class containing atomics and closed under boolean operations and *restricted* quantifiers. Thereafter Π_{n+1}^{Levy} (resp. Σ_{n+1}^{Levy}) formulae are obtained from Σ_n^{Levy} (resp. Π_n^{Levy}) formulae by binding some or all the free variables with $\forall$ (resp. $\exists$). In Levy [1965] the formulae usually have a theory superscript. This is because we wish to say that a formula is Δ_n^T if it is T-equivalent to both a Σ_n^{Levy} and a Π_n^{Levy} formula. However, here there will be so few uses of this terminology that no such disambiguation will be required.

The following classification of formulae is of particular interest here. A $\Delta_0^{\mathcal{P}}$ formula is a member of the smallest class containing atomics and closed under the propositional connectives, and the usual restricted quantifiers $Qx \in y \ldots$, and a new style of restricted quantifier $Qx \subseteq y \ldots$. (The "$\mathcal{P}$" stands for "$\mathcal{P}$ower set".) Thereafter we define $\Pi_{n+1}^{\mathcal{P}}$ to be the closure of $\Sigma_n^{\mathcal{P}}$ under $\forall$. $\Sigma_{n+1}^{\mathcal{P}}$ is dually the closure of $\Pi_n^{\mathcal{P}}$ under $\exists$. The reason for interest in this hierarchy is that $\Sigma_1^{\mathcal{P}}$ sentences are preserved by end-extensions for which the power set operation is absolute. End-extensions of models of set theory with $V \in V$ will preserve power set as long as (the old) V is a power set in (the new) V. Since this happens in several natural constructions of end-extensions below we shall find ourselves thinking of $\Sigma_1^{\mathcal{P}}$ sentences frequently. $i : \mathcal{A} \hookrightarrow_e^{\mathcal{P}} \mathcal{B}$ ("i is a $\mathcal{P}$-embedding from $\mathcal{A}$ into $\mathcal{B}$") means that $i : \mathcal{A} \to \mathcal{B}$ is an embedding for which the power set operation is absolute. "No new members or subsets of old sets." If i is the identity we say $\mathcal{B}$ is a **$\mathcal{P}$-extension** of $\mathcal{A}$. For more on $\mathcal{P}$-extensions see Forster and Kaye [1991] and references therein. We shall write "$M \subseteq_e^{\mathcal{P}} N$" if N is an end-extension of M with the same power set relation similarly.

In any set theory with a universal set, the Levy hierarchy and the $\mathcal{P}$ hierarchy collapse: every formula is Δ_2^{Levy}. If Φ is an arbitrary formula, restrict all variables in it to a new variable x (getting Φ^x which is Δ_0^{Levy}) and then Φ is equivalent to both

$$\forall x \exists y (y \notin x \vee \Phi^x)$$

and

$$\exists x \forall y (y \in x \ \wedge \ \Phi^x).$$

Indeed, if we are willing to use lots of x, so that we add one new variable for each type in ϕ, then we can ensure that each stratified wff is equivalent to a stratified Δ_2^{Levy} sentence. We shall find ourselves making much more use of the $\mathcal{P}$ hierarchy than the Levy hierarchy.

1.1.2 *Mainly concerning type theory*

There are two type-theoretic languages that will concern us here, both going back to Russell [1908]. TST, *Simple Type Theory*, is expressed in a language with a type for each non-negative integer, an equality relation at each type, and between each pair of consecutive types n and $n+1$ a relation $\in_n$. The axioms are an axiom of extensionality at each type

$$\forall x_{n+1} \forall y_{n+1} \ (x_{n+1} = y_{n+1} \longleftrightarrow \forall z_n (z_n \in_n x_{n+1} \longleftrightarrow z_n \in_n y_{n+1}))$$

and (at each type) an axiom scheme of comprehension

$$\forall \vec{x} \ \exists y_{n+1} \forall z_n \ (z_n \in_n y_{n+1} \longleftrightarrow \phi(\vec{x}, z_n))$$

with 'y_{n+1}' not free in 'ϕ'.

TST_k is like TST except that there are only k types, labelled 0,..., $k-1$. The "theory of negative types"[5] (*TNT*) (Wang [1952a]) and its language are defined analogously, except that the types are indexed by Z.

[5]This is actually a misnomer: strictly he should have called it the theory of *positive and* negative types.

TSTI is TST with the axiom of infinity for the bottom level. (We have to make explicit that we mean the bottom level, for it is possible to have a Dedekind-finite set whose power set has a countably infinite subset, so we can have models of type theory which are infinite above some level but Dedekind-finite below it.) Similarly we will have $TSTI_k$. In general, the result of appending an '*I*' to the name of a theory denotes the result of adding the axiom of infinity to that theory. If x is a set in a model of some minimal sensible set theory (Zermelo set theory will do), $\langle\langle x\rangle\rangle$ is the structure $(x, \mathcal{P}`x, \mathcal{P}^2`x, \ldots)$ thought of as a model of TST.

Readers for whom the expression 'type theory' connotes Church [1940] should be warned by this that we are here dealing with something much simpler!

Suppose $M \models$ TST. Form a new model M^* by deleting the bottom type (type 0) from M and relabelling the old type 1 as type 0, with the others consequently. ϕ^+ is the result of raising all type indices in ϕ by 1, and ϕ^n is the result of raising all type indices in ϕ by n.

A formula of set theory is *stratified* iff by assigning type subscripts to its variables we can turn it into a wff (well-formed formula) of simple type theory. That is to say, a wff ϕ is stratified iff we can find a *stratification assignment* (henceforth "stratification" for short) for it, namely a map f from its variables (after relettering where appropriate) to $\mathbb{N}$ such that if the atomic wff '$x = y$' occurs in ϕ then $f(`x\text{'}) = f(`y\text{'})$, and if '$x \in y$' occurs in ϕ then $f(`y\text{'}) = f(`x\text{'}) + 1$. Variables receiving the same integer in a stratification are said to be of the same *type*. If n successive integers are used, the formula is said to be *n-stratified*. There is a notion of a *canonical stratification* which assigns each variable the lowest possible type. A formula with one free variable, and that being assigned type n in the canonical stratification, is an *n-formula*. If Φ and Ψ are two closed stratified formulae, then we can assign integers to their variables independently, and so the canonical stratification for $\ulcorner\Psi \wedge \Phi\urcorner$ will be that function whose restrictions to the two sets of variables are the two canonical stratifications. If ϕ is a stratified formula then $\ulcorner\#\phi\urcorner$ is that expression in the language of type theory which is the result of incorporating into ϕ as type subscripts the integers used in the canonical stratification.

The notion of a stratified formula is much more natural than one might think at first. There is a theorem of Coret [1970] that every stratified instance of the axiom scheme of replacement is provable in Z, Zermelo set theory, and a theorem of Boffa and Mathias that $V_{\omega+\omega} \prec_{\text{strat}} H_{\beth_\omega}$ in ZF. ("$\prec_{\text{strat}}$" means "is a substructure elementary for stratified formulae", and H_α is defined in section 1.1.3 below.) The notion is also intimately connected with that of a well-typed formula of polymorphic λ-calculus. The n-stratified Δ_0^{Levy} formulae express concepts that are simple in the sense that to find out whether or not x bears some n-stratified property ϕ it is sufficient to examine members x, $\bigcup x$,..., $\bigcup^{(n-1)} x$ only, rather than the whole of $TC(x)$. This impression of a complexity hierarchy will be reinforced by the appearance in higher levels of truth definitions for formulae belonging to lower levels (see McNaughton [1953]). Finally there is a preservation theorem in chapter 3 to the effect that in the language of set theory the stratified sentences are precisely those preserved by a particular model-theoretic construction to be revealed in that chapter.

If $\Phi(\vec{x})$ is a formula where such failures of stratification as there are involve the free variables only—that is to say, we can give subscripts to the bound variables which obey the stratification rules—then Φ is said to be *weakly stratified*. Weakly stratified formulae are obtained from stratified formulae by identifying free variables. The significance of weak stratification is that, if we wish to set up *NF* as a natural deduction system, we will need $\in$-introduction and elimination rules for weakly stratified formulae. A stratified formula Φ is said to be *homogeneous* iff all its free variables are of the same type.

An expression $\Phi(x)$ (in one free variable 'x') is *typed* if there is a closed stratified Ψ so that $\Phi(x)$ is the result of replacing all quantifiers Qy in Ψ by the restricted quantifier $Qy \in \mathcal{P}^n\text{`}x$, where n is the type 'y' receives in some fixed stratification. All typed formulae are stratified $\Delta_1^{\mathcal{P}}$ formulae with one free variable.

To discuss these topics properly we will also need $j =_{\text{df}} \lambda f \lambda x.(f\text{“}x)$. The map j commutes with group-theoretic operations: $j\text{`}(\pi\sigma) = j\text{`}\pi j\text{`}\sigma$. Since we are dealing with set theory with $V \in V$, the full (internal) symmetric group on the universe may well turn out to be a set too, and we will write 'J_0' for it. J_{n+1} is $j\text{“}J_n$. The notation “J_n” is intended to suggest to the reader that it denotes the group of those permutations that are at least j^n of something. It will turn out (modulo some existence axioms) that if Φ is an n-formula, then $(\forall x)(\forall \tau \in J_n)(\Phi(x) \longleftrightarrow \Phi(\tau\text{`}x))$. If $\exists \sigma \in J_n\ \sigma\text{`}x = y$ then we will say that x and y are *n-equivalent* and write this '$x \sim_n y$'. A set x is *n-symmetric* iff, for all permutations τ, $(j^n\text{`}\tau)\text{`}x = x$ and is *symmetric* iff for some n it is n-symmetric. In *NF* all stratified set abstracts are symmetric.[6]

A (possibly external) permutation σ of a set X is *setlike* if, for all n, $j^n\text{`}\sigma$ is defined and is a permutation of $\mathcal{P}^n\text{`}X$. If we have a (*GB*-style) axiom scheme of replacement then it is trivial that any permutation is setlike, but in general it is not obvious that $j\text{`}\sigma$ is onto. An outer automorphism of $\langle V, \in \rangle$ would be an example of a setlike permutation of V that is not a set.

[6]This is also true for all sets definable by stratified expressions, because the unique thing that is ϕ is $\{x : \exists z(\phi(z) \wedge x \in z)\}$.

1.1.3 *Other definitions*

Cardinals are notated with a superscripted '=' thus: $\overline{\overline{x}}$. We will sometimes write '$x \sim y$' instead of '$\overline{\overline{x}} = \overline{\overline{y}}$'.

H_ϕ is the collection of all x s.t. $(\forall y)(\{z \subseteq y : \phi(z)\} \in y \to x \in y)$. A few minutes with pencil and paper should enable the reader to discover that this object is the least fixed point for $\lambda x.\{y \subseteq x : \phi(y)\}$, and the '$H$' connotes 'hereditarily'. (Sometimes we will be interested in the maximal fixed point, but we do not have a notation for this.) We will write 'H_α' for '$H_{\overline{\overline{x}}<\alpha}$' (in plain language: the things hereditarily smaller than α) and 'H_X' for the things hereditarily in X.

Σ_X is the full symmetric group on a set X.

A dash '$-$' is set complementation when monadic and set difference when dyadic. $x \Delta y$ is the symmetric difference of x and y: $(x \cup y) - (x \cap y)$.

$B`x = \{y : x \in y\}$.[7] A *Boffa atom* is a fixed point for this operation.

$F`x$ ('F' for 'filter') is the set of supersets of x.

$b`x$ is the set of things which meet $x$. Thus $b`x = -\mathcal{P}` - x$. The '$b$' is an upside-down '$\mathcal{P}$' to remind us that these operations are dual.

Most of the *NF*-like theories have an axiom giving us $B`x$, that is $\{y : x \in y\}$. This is an enormously important operation. It is intimately involved in the proofs that $NF = NF_4$ (proposition 2.3.12); that every model of *NF* has a proper end-extension partly describable in the ground model (theorem 3.1.13); that every countable binary structure can be embedded in every model of *NFO* (theorem 2.4.3: see definition of this theory in section 1.1.4); and that *NF* is finitely axiomatizable (page 26) and has no $\subseteq$-minimal model (remark 2.1.11). It is worth noting that if X is a finite set then $\mathcal{P}^2`X$ is the free boolean algebra generated by $B``X$. Church and Mitchell's systems do not have this axiom and proofs of analogous results for them are either difficult or impossible.

It is common to use 'ι' to denote the singleton function: thus "$\iota`x$" instead of "$\{x\}$". This is partly to avoid confusion with '$\{x : \Phi\}$' and partly because, particularly in *NF*, the operation sending x to $\iota``x$ is quite important, and would be hard to notate without 'ι'. Rosser [1953a] writes this function as "$USC(x)$" but the ι notation is older (it is in Peano) and more economical. Rosser also has the notation "$RUSC(R)$" for 'R^ι', $\{\langle \iota`x, \iota`y\rangle : \langle x, y\rangle \in R\}$. In this case we retain Rosser's notation instead of 'R^ι' because it is deeply engrained in *NF* practice and because the 'R^ι' suggests to the eye that ι is a set which (in *NF* at least) it is not. An $x = \iota`x$ is a *Quine atom.* (Forti and Honsell call them *autosingletons.*)

A *set abstract* or *term* of a (set) theory T is $\{x : \Phi\}$, where 'x' is the

[7]This function was first considered by Quine, who used a different notation for it. Whitehead suggested to him that it should be called the "essence" of x.

sole free variable in Φ and the existence of $\{x : \Phi\}$ is an axiom of T.[8] The *term model* for a complete theory T is the family of all set abstracts of T with membership relation $t_i \in t_j$ iff $T \vdash t_i \in t_j$, and equality $t_i = t_j$ if $T \vdash t_i = t_j$. A term model for an incomplete theory T is the term model of some complete extension of T. A (or the) term model for T may or may not be a model of T. Some theories (*NFO* and NF_2), which are not complete, nevertheless decide all equalities and membership questions about their terms, and so we can speak of *the* term model for *NFO* and NF_2. They satisfy *NFO* and NF_2 respectively. In contrast, it is unknown if any term model for *NF* satisfies it.

1.1.4 *Theories*

We think of theories here as inductively defined by an axiomatization under deductive closure. Thus we might consider different presentations of the same set of sentences to be distinct theories. (See the last paragraph of the previous section.) If T is a theory, *str*(T) is the theory whose axioms are the stratified theorems of T (the notation is Orey's [1964]). *NF* is extensionality plus the axiom scheme giving the existence of $\{x : \phi(x, \vec{y})\}$ for ϕ stratified. N_kF is that subtheory of *NF* consisting of extensionality plus the existence of $\{x : \phi(x, \vec{y})\}$ where $\phi(x, \vec{y})$ can be stratified using no more than k types ($\phi(x, \vec{y})$ may contain free variables). NF_k is the subsystem with extensionality plus such comprehension axioms as can be stratified using no more than k types. The difference can be illustrated with an example. '$\forall x \exists y \forall z\ (z \in y \longleftrightarrow x \in z)$' is an axiom of NF_3 (since any stratification uses only 3 integers) and of N_2F (since '$x \in z$' is stratifiable with only two integers) but it is not an axiom of NF_2. Several of these theories will be important to us: N_1F is the axiom scheme saying, for each n, that there are at least n distinct objects. A model of NF_2 is simply an infinite atomic boolean algebra with the same number of atoms as elements. Accordingly NF_2 is finitely axiomatizable as extensionality plus existence of $\iota`x$, $x \cup y$, $x \cap y$, and $-x$. N_1F, NF_2, N_2F, and NF_3, are all distinct. Since I shall not be greatly concerned with most of these systems here, the interested reader should consult Oswald (all *op. cit.*) for proofs of this fact.

There are subtheories which do not fit into this classification scheme. If Γ is a set of stratified formulae, then $NF\Gamma$ is extensionality plus all instances of '$\forall \vec{x} \exists y \forall z\ (z \in y \longleftrightarrow \phi)$', where $\phi \in \Gamma$ (with 'y' not free in ϕ). Uses of this notation when Γ is not a set of formulae include: *NFO* is extensionality plus existence of $\{x : \Phi\}$ with Φ stratified and quantifier free (*NFO* is in fact NF_2 plus existence of $B`x$); *NFU* is *NF* with extensionality weakened

[8] 'Axiom of T' is not an oversight. The intention of this definition is to prevent us from being obliged to include accidental unstratified set abstracts in term models for *NF*.

to hold for non-empty sets only (*TSTU* and *TNTU* are the corresponding type theories); later we will define a system known as '*NFC*' which is *not* *NF* + *AC*! *CUS* is Church's universal set theory, which will be defined in chapter 4. *ZF* is *ZF* (i.e. without *AC*); *Z* is Zermelo's set theory, likewise without choice.

1.2 Some motivations and axioms

We can discern two traditions in set theory, and they can be roughly associated with Zermelo and Russell. There is a tradition (in which we find *ZF*) of thinking of set theory as a branch of mathematics, so that sets are mathematical objects. In this tradition we also find the systems *CUS* of Church, and of Mitchell, and the free construction principles of Forti and Honsell [1983, 1984a]. The other tradition, sets as allegories, is associated more with logicism and type theory. I shall concentrate on set membership as an allegory for predication.

1.2.1 *Sets as predicates-in-extension*

Sets are predicates-in-extension. The use of the idiom 'isa' by modern workers in artificial intelligence is an explicit acknowledgement of this, as is the use of the letter '$\in$' in more formal contexts. Peano introduced the epsilon for this purpose since it is the first letter of the Greek word that functions as the 'is' of predication.[9]

Someone once said to me, "Doing set theory with a universal set is like believing in God." This arresting simile expresses a sound intuition that, even though the distinctions suggested by the limitative theorems are reflected nicely in type theory and ZFC, these distinctions take no cognizance of the fact that our *consciousness* and our *experience* are unitary. We may sensibly attach ordinal subscripts to synthetic languages or mathematical objects: we do not attach them to ourselves, nor to our thoughts, and certainly not to God. To use a medical metaphor, set membership *presents* as an allegory of predication, but it is also an allegory in more obscure ways, for example of the dominance relations of the numerous conceptual hierarchies that people have dreamed up from time to time, be they of language levels or levels of existence (humans, ..., angels, ..., God). In view of this, we find ourselves in need of an approach to set theory that does not dismiss these intuitions as mere folk psychology. Somewhere we have to "restore severed connections" as Quine puts it. I am not pretending that I have even expressed the problem well, let alone that the approach I am most familiar with (*NF*) holds the key to the answer, merely that it is a problem we should take seriously and that, with a view of $\in$ as allegory, we should explore set theory with a universal set in a serious and constructive

[9]See, for example, Quine [1986] and other articles in that volume.

spirit. We should not expect that the insights it will have to offer will come easily. Having found such an approach, we will need to examine very closely anything it has to say about truth definitions.[10]

The idea that $\in$ is an allegory for predication is deeply rooted in the logicist tradition.[11] Its continuing fertility is attested by such phenomena as the emergence of situation semantics (Barwise [1984] and references therein). Set theory is a branch of *logic*, the logic of *predication*. The allegory carries with it no justification for the axiom of foundation, since V is the extension of the predicate of self-identity and will clearly be self-membered. This point of view does not commit us to extensionality (though we may have other reasons for insisting on it), and in particular it certainly does not exclude *urelemente*, even if we wish to retain extensionality for non-empty structures. It is significant that *NFU* (*NF* with *urelemente*) is equiconsistent with a weak fragment of arithmetic, whereas *NF* itself is quite possibly at least as strong as Z: it certainly proves the axiom of infinity.

One can argue for extensionality in this context by taking a more specific view than this more general one. Sets are properties in *extension.* It is lexically obvious that objects that are extensions of something should satisfy a form of extensionality.

1.2.1.1 *Some axioms* One axiom this naturally leads us to is complementation: $-x$ is a set if x is a set, and all the theories we examine in this book have such an axiom (though topological set theory does not: see section 1.2.2). The idea that there is a kind of duality between large sets and small (the allegory of negation) can be extended if we imagine an operation $\hat{}$ on formulae such that $\hat{\phi}$ is the result of replacing all occurrences of $\in$ in ϕ by $\notin$ and vice versa. We can then consider schemes like "$\Phi \longleftrightarrow \hat{\Phi}$", and an extreme version, postulating the existence of *antimorphisms*, permutations σ of V so that $\forall xy(x \in y \longleftrightarrow \sigma`x \notin \sigma`y)$ (see Forster [1985]).

Relevant to this logicist view of $\in$ as predication is the involvement of individual logicists (Russell, Quine) in the normative rôle logic has played in twentieth century Anglo-Saxon philosophy: to wit, a way of weeding out rubbish. If we combine the logicist view of $\in$ with a view of logic as bush-clearing, we have a point of view that set theory should axiomatize an abstract theory of predication while introducing in the process as few nasty sets as possible. This view will never tell us that $\in$ is well-founded of course, but nor will it lead us to suppose that there are, for example, sets $x = \iota`x$. Such a set cannot be the extension of any sensible predicate.[12]

[10] For more in this vein see Kripke [1975].

[11] See also McDermott [1977].

[12] By a nice irony such sets are now called "Quine atoms" for Quine, working in this logicist tradition in [1967], suggested that when y is an *urelement*, "$x \in y$" could be read

It is fitting, even if coincidental, that of all the systems we will consider, Quine's *NF* lends itself best to this kind of development: for example, it is easy to show that *NF* (if consistent) does not prove the existence of any $x = \iota`x$. In contrast to the Forti–Honsell/Aczel point of view, which gives rise to *anti*foundation axioms, this gives rise to *quasi*foundation axioms. We will now consider some of them briefly.

∈-determinacy and pseudoinduction

In this section we work in naïve set theory, not because we wish to prove things in inconsistent theories, but because we are more interested in speculation than formal development.

If we are to work in a set theory with a universal set, we must of course abandon well-foundedness of ∈ and the principle of ∈-induction that goes with it. Are there any induction principles we can hang on to? There are, and, pleasingly, they are best approached through games.

However we will start by setting these ideas in a more general context, without the specific restriction to ∈ which will be our local motivation.

Fix a binary structure $\langle X, R\rangle$ for the moment. Consider the game, notated 'G_x' for the nonce, played by starting with an element $x_0 \in X$ in which players I and II alternately (I starting) pick $\{x_1, x_2, x_3, \ldots, x_n \ldots\}$ all in X with $x_{n+1} R x_n$ for all n, and I picking elements with odd subscripts and II picking elements with even subscripts. The game continues until one player attempts to find x_{n+1} when x_n is R-minimal and thereby loses. If the game goes on for ever it is a draw. If R is well-founded all such games over X are of course determinate but the converse is not true. This pseudo well-foundedness might turn out to retain some of the desirable properties of well-foundedness, such as giving rise to induction principles.

Let us adopt the following definitions for the moment. (We will later want to use 'I' and 'II' in subtly different ways.)

$$\mathrm{I}_R =_{\mathrm{df}} \{x : \mathrm{I} \text{ Wins } G_x\}; \ \mathrm{II}_R =_{\mathrm{df}} \{x : \mathrm{II} \text{ Wins } G_x\}.$$

More generally, we could adopt the following definition.

DEFINITION 1.2.1 *$R \subseteq X \times X$ is pseudowellfounded iff$_{df}$ X can be partitioned into $\mathrm{I}_R \cup \mathrm{II}_R$ such that for all $X' \subseteq X$ we have*

$$(((\forall x)(\forall yRx)(\exists zRy)(z \in X') \to x \in X') \to \mathrm{II}_R \subseteq X')$$

and

$$(((\forall x)(\exists yRx)(\forall zRy)(z \in X') \to x \in X') \to \mathrm{I}_R \subseteq X').$$

Just as we can extract, from any structure $\langle X, R\rangle$ whatever, the well-founded part of X by means of a transfinite construction mimicking the

to be the same as "$x = y$" on the grounds of notational economy.

cumulative hierarchy of sets, so we can extract, from any structure $\langle X, R\rangle$ whatever, the pseudowellfounded part of X by means of a transfinite construction mimicking the construction of $\mathtt{I}_\alpha$ and $\mathtt{II}_\alpha$. This pseudowellfounded part is naturally partitioned into two sets we can call $\mathtt{I}_R$ and $\mathtt{II}_R$. This of course justifies the twin principles of pseudoinduction:

$$\frac{(\forall yRx)(\exists zRy)(\phi(z)) \to \phi(x)}{(\forall x \in \mathtt{II}_R)(\phi(x))} \qquad \frac{(\exists yRx)(\forall zRy)(\phi(z)) \to \phi(x)}{(\forall x \in \mathtt{I}_R)(\phi(x))}$$

for any expression ϕ such that our enveloping theory proves that $\{z : \phi(z)\}$ is a set. These are **I-induction** and **II-induction**.

Of course we have I-induction and II-induction anyway (just as we have R-induction for the well-founded part of $\langle X, R\rangle$ anyway): pseudowellfoundedness is the additional assumption $X = \mathtt{I}_R \cup \mathtt{II}_R$.

We must work this backwards to show that the condition on $\langle X, R\rangle$ that we need in order to justify pseudoinduction over $\langle X, R\rangle$ is precisely that $\langle X, R\rangle$ should be pseudowellfounded, just as well-foundedness of R is precisely what we need to justify R-induction.

We derive well-foundedness by considering what conditions we have to put on R to justify R-induction.

Let us do the same for R-pseudoinduction (over a set X). We have partitioned X, the domain of R, into two pieces, $\mathtt{I}_R$ and $\mathtt{II}_R$, and we want to find what conditions have to be put on $\langle X, R\rangle$ to justify I-induction and II-induction over R.

Suppose this inference

$$\frac{(\forall yRx)(\exists zRy)(\phi(z)) \to \phi(x)}{(\forall x \in \mathtt{II}_R)(\phi(x))}$$

fails. Then we know that there is a non-empty set X' (namely the set of things that aren't ϕ) such that $(\forall y \in X')(\exists xRy)(\forall x'Rx)(x' \in X')$ (because if not, we must have $\phi(y)$). We also know that X' meets $\mathtt{II}_R$.

Similarly suppose the inference

$$\frac{(\exists yRx)(\forall zRy)(\phi(z)) \to \phi(x)}{(\forall x \in \mathtt{I}_R)(\phi(x))}$$

fails. Then we know that there is a non-empty set X' (namely the set of things that aren't ϕ) such that $(\forall y \in X')(\forall xRy)(\exists x'Rx)(x' \in X')$ (because if not, we must have $\phi(y)$). We also know that X' meets $\mathtt{I}_R$.

R-pseudoinduction holds if these two things cannot happen. These cannot happen if

$$(\forall Y \subseteq X)(Y \cap \mathtt{I}_R \neq \Lambda \to (\exists x \in Y)(\exists yRx)(\forall zRy)(z \notin Y))$$

and

$$(\forall Y \subseteq X)(Y \cap \mathtt{II}_R \neq \Lambda \rightarrow (\exists x \in Y)(\forall y R x)(\forall z R y)(z \notin Y)),$$

which of course is the condition of pseudowellfoundedness.

In set theory the concept of a **regular** set arises from the desire to deduce well-foundedness of $\in$ from $\in$-induction. One proves by $\in$-induction on x that any set to which x belongs is disjoint from one of its members. This property of x is **regularity**. In this context we prove analogously by I-induction on x that

$$(\forall X' \subseteq X)(x \in X' \rightarrow (\exists y \in X')(\exists w R y)(\forall z R w)(z \notin X'))$$

and prove by II-induction on x that

$$(\forall X' \subseteq X)(x \in X' \rightarrow (\exists y \in X')(\forall w R y)(\exists z R w)(z \notin X')).$$

Accordingly we might wish to consider an axiom to the effect that $\in$ is pseudowellfounded. This is the axiom of $\in$-pseudofoundation, which is either of the following equivalent assertions:

1. $\in$ is pseudowellfounded;
2. $(\forall x)(G_x$ is determinate).

We shall write 'I' and 'II' instead of '$\mathtt{I}_\in$' and '$\mathtt{II}_\in$'

The first thing to do is to tie up a few loose ends and check that we can use I-induction to prove that $(\forall x \in \mathtt{I})(\mathtt{I}$ Wins $G_x)$ and II-induction to prove that $(\forall x \in \mathtt{II})(\mathtt{II}$ Wins $G_x)$. This is easy enough to be left as an exercise.

We shall see later that term models for *NF* (if there are any) are models for pseudofoundation for stratified formulae, and that if *NF* is consistent it has a model where V is the disjoint union of two *sets* X and Y with $X = \mathcal{P}`Y$ and $Y = b`X$.

Obviously $x \in \mathtt{I}$ iff $(\exists y \in x)(y \in \mathtt{II})$, and dually $x \in \mathtt{II}$ iff $(\forall y \in x)(y \in \mathtt{I})$. Using '$b$' as defined in section 1.1, we can note that this implies

$$\mathtt{I} = b`\mathtt{II}, \ \ \mathtt{II} = \mathcal{P}`\mathtt{I}.$$

Naturally we are going to want $\langle \mathtt{I}, \mathtt{II}\rangle$ to be the least fixed point for the function: $\lambda x.\langle b`(\mathtt{snd}(x)), \mathcal{P}`(\mathtt{fst}(x))\rangle$. Inductive definitions always come in two matching flavours: "from above" and "from below". Clearly the correct definition of I and II "from above" is going to be

$$\mathtt{I} \ = \bigcap\{y : b`\mathcal{P}`y \subseteq y\} \text{ and } \mathtt{II} \ = \bigcap\{y : \mathcal{P}`b`y \subseteq y\}.$$

To every recursive definition of a class there corresponds of course an induction principle. The definitions of I and II "from above" give rise immediately to I-**induction** and II-**induction**.

For the correct definition "from below" we notice first that $\Lambda \in \mathrm{II}$ and $V \in \mathrm{I}$, so we naturally find

$$\mathrm{I}_0 = \iota\text{`}V, \quad \mathrm{I}_{\alpha+1} = b\text{`}\mathrm{II}_\alpha,$$

$$\mathrm{II}_0 = \iota\text{`}\Lambda, \quad \mathrm{II}_{\alpha+1} = \mathcal{P}\text{`}\mathrm{I}_\alpha,$$

taking sumsets at limit ordinals. (Compare with the definition of the cumulative hierarchy: $V_0 = \Lambda$, $V_{\alpha+1} = \mathcal{P}\text{`}V_\alpha$, taking sums at limits.) It is perhaps worth noting that

REMARK 1.2.2 *I_α and II_α are increasing sequences under $\subseteq$.*

Proof: We prove by induction on α that $(\forall\beta < \alpha)(\mathrm{I}_\beta \subseteq \mathrm{I}_\alpha)$ and $(\forall\beta < \alpha)(\mathrm{II}_\beta \subseteq \mathrm{II}_\alpha)$. For successor cases we reason as follows

$$\mathrm{I}_\alpha \subseteq \mathrm{I}_{\alpha+1} \wedge \mathrm{II}_\alpha \subseteq \mathrm{II}_{\alpha+1}$$

but b and $\mathcal{P}$ are both increasing so we have

$$\mathcal{P}\text{`}\mathrm{I}_\alpha \subseteq \mathcal{P}\text{`}\mathrm{I}_{\alpha+1} \wedge b\text{`}\mathrm{II}_\alpha \subseteq b\text{`}\mathrm{II}_{\alpha+1}$$

and

$$b\text{`}\mathcal{P}\text{`}\mathrm{I}_\alpha \subseteq b\text{`}\mathcal{P}\text{`}\mathrm{I}_{\alpha+1} \wedge \mathcal{P}\text{`}b\text{`}\mathrm{II}_\alpha \subseteq \mathcal{P}\text{`}b\text{`}\mathrm{II}_{\alpha+1},$$

which is to say

$$\mathrm{I}_{\alpha+1} \subseteq \mathrm{I}_{\alpha+2} \wedge \mathrm{II}_{\alpha+1} \subseteq \mathrm{II}_{\alpha+2}.$$

For λ a limit we need to know that

$$\mathrm{I}_\lambda \subseteq \mathrm{I}_{\lambda+1} \wedge \mathrm{II}_\lambda \subseteq \mathrm{II}_{\lambda+1}.$$

Now

$$\mathrm{I}_\lambda = \bigcup_{\beta<\lambda} \mathrm{I}_\beta \wedge \mathrm{II}_\lambda = \bigcup_{\beta<\lambda} \mathrm{II}_\beta.$$

So $(\forall\beta < \lambda)(\mathrm{I}_\beta \subseteq \mathrm{I}_\lambda)$, whence

$$(\forall\beta < \lambda)(b\text{`}\mathcal{P}\text{`}\mathrm{I}_\beta \subseteq b\text{`}\mathcal{P}\text{`}\mathrm{I}_\lambda = \mathrm{I}_{\lambda+1}),$$

which implies that

$$\mathrm{I}_\lambda \subseteq \mathrm{I}_{\lambda+1}.$$

and II similarly. ■

DEFINITION 1.2.3 *The pseudorank of a set x is the least β such that it belongs to I_β or to II_β.*

I-induction and II-induction imply that everything in I or II has a pseudorank.

The player with the winning strategy has a particularly nice strategy because (s)he is trying to get into an open[13] set and, in those circumstances, arbitrary unions of Winning non-deterministic strategies are Winning non-deterministic strategies (this is *not* true in general for arbitrary games) so they can play

The lazy strategy:

"When confronted with x, play anything in $x \cap \texttt{II}$ of minimal rank. If $x \cap \texttt{II}$ is empty, do anything at all."

It is standard that the rank of a well-founded set x can be defined *either* as the rank of $\in$ restricted to $TC(x)$ (considered as a well-founded relation) *or* as the least ordinal α such that $x \in V_\alpha$. There is a corresponding result here.

THEOREM 1.2.4 The pseudorank theorem

The pseudorank of x is the same as the rank of the tree of plays obtainable in G_x by the winning player using his/her lazy strategy and the other player doing all possible things.

Proof: We prove this by induction on pseudorank.

The reader may also note that rank and pseudorank agree on well-founded sets.

$\langle \texttt{I}, \texttt{II} \rangle$ is the least fixed point for $\lambda x.\langle b`(\texttt{snd}(x)), \mathcal{P}`(\texttt{fst}(x))\rangle$. It might be an idea to look at what the family of fixed points for this function must look like.

Now we can return to the question of the sethood of I and II. Mirimanoff's paradox tells us that WF cannot be a set, and there is an analogous result here. In fact there are two proofs that I and II cannot be sets, but one of them is not very robust. We begin with it, since the second one, although more elementary, and quite analogous to the derivation of Mirimanoff's paradox, is less pleasing.

First we show that every set in I or II has a pseudorank. We saw this in theorem 1.2.4. It is clear that I, if a set, is a member of II. What is its pseudorank? By theorem 1.2.4 its pseudorank is at least as great as the pseudorank of anything in I. But the pseudorank of $\{\texttt{I}\}$, which is certainly a member of I, must be one more than the pseudorank of I, which is impossible. ■

Notice that this does not use $\in$-pseudofoundation! The other proof that I and II cannot be sets is an immediate corollary of the following remark.

REMARK 1.2.5 *There is no $\subseteq$-least set x such that $\mathcal{P}`b`x \subseteq x$.*

[13]We give V the discrete topology and V^ω the product topology. The history of the play is considered to be an element of V^ω.

Proof: Suppose $\mathcal{P}`b`x \subseteq x$. Note that in this case $b`x \in x$. We will show that $\mathcal{P}`b`(x - \{b`x\}) \subseteq (x - \{b`x\})$.

Now $\mathcal{P}`b`(x - \{b`x\}) \subseteq x$ so it will suffice to show that $w \subseteq b`(x - \{b`x\}) \rightarrow w \neq b`x$. To prove this it will suffice to show that $b`x \not\subseteq b`(x - \{b`x\})$. But if $b`x \subseteq b`(x - \{b`x\})$ we must have $x \subseteq (x - \{b`x\})$, which we noted at the outset is not the case. ■

In particular if $x = \bigcap\{y : \mathcal{P}`b`y \subseteq y\}$ then $b`x$ both is and is not a member of $x$. This is analogous to Mirimanoff's paradox that $\bigcap\{y : \mathcal{P}`y \subseteq y\}$ both is and is not a member of itself. (This result is analogous to remark 2.1.23)

We note at this stage (although we will not prove it yet) that it is consistent relative to *NF* that $\lambda x.\langle b`(\texttt{snd}(x)), \mathcal{P}`(\texttt{fst}(x))\rangle$ should have a fixed point whose two components are complements.

There is an important philosophical contrast between the axiom of foundation and an axiom of pseudofoundation. It is possible to persuade oneself that the von Neumann construction is indeed a construction (so that we can pretend we don't have to be platonists but can be engineers instead and be like God. [After all God is an engineer not a mathematician: she is a *creative* force not a *knowing* force.]) It is certainly *not* possible to pretend that the recursive construction of $\texttt{I}$ and $\texttt{II}$ is anything other than a selection of pre-existing objects. If you want to believe $V = \texttt{I} \cup \texttt{II}$ you will have the devil of a job explaining how V, which is the end result of your construction, is something you just happened to rustle up at the second stage of your construction, since $V \in \texttt{I}_2$.

Strong extensionality

All there is to know about a set is its members. This is not so much a feature peculiar to sets (though it is that), as one utterly fundamental to our conception of them. One can do set theory without the axiom of choice; or without the axiom of foundation; or without the axiom of power set; or without complicated replacement—all of these have been done—but one cannot do set theory without the axiom of extensionality for non-empty sets. This view of what is central to set theory is not tied to a logicist or to a sets-as-predicates view: it is much more widespread and less controversial.

If we have the axioms of foundation and extensionality, then the question of when two sets are identical has an elegant recursive solution. If we know what it is for two *urelemente* to be identical, then well-foundedness of $\in$ tells us that the obvious transfinite algorithm will always terminate. Because of this, people brought up on set theories with the axiom of foundation have always been able to avoid thinking about the question of equality between sets. What can we say in the absence of the axiom of foundation? If we do not decide what set equality is to be on the basis of some robust philosophical principle then we are implicitly accepting a plurality of answers, so that set theories become like groups, or topological spaces, and we are in effect treating equality as just another predicate letter. This is repugnant to a long tradition of treating equality as a logical connective.

The only principle with anything overt to say on the matter is extensionality, and that will tell us only that set equality is an equivalence relation $\sim$ on V so that

$$\forall xyz(x \sim y \land z \in x \to (\exists w)(z \sim w \in y)).$$

Such equivalence relations are considered by Hinnion [1980, 1981] (who calls them *contractions*) and also by Forti and Honsell (see the bibliography). Thus $\sim$ is a contraction iff it is a fixed point for the operation which Hinnion calls '+', where

DEFINITION 1.2.6

$$x \sim^{+} y \longleftrightarrow_{\text{df}} (\forall z \in x)(\exists w \in y)(z \sim w) \land (\forall z \in y)(\exists w \in x)(z \sim w).$$

Relations like this crop up in theoretical computer science, where they are called *bisimulations.* One can think of the $\in$-diagram of a set as the diagram of the set of states of a computation. Two states $\sim$ each other if they *simulate* each other. This is widely discussed in the literature of theoretical computer science, but the most appropriate reference here is probably Aczel [1988]. The principle that *sets are that which is extensional* will tell us only that $=$ is a contraction or bisimulation, but we can appeal to the logicist "minimal rubbish" assumption to tell us there is a unique contraction and thus a canonical answer. If there is to be only one contraction, it will be equality; as it happens, this enables us to characterize identity by means of a game, the "identity" game, $G_{x=y}$, to which we now turn.

There are two players, $=$ and $\neq$. The idea is that $=$ is trying to prove that $x = y$ and $\neq$ is trying to prove that $x \neq y$. Player $=$ moves first, choosing $R_1 \subseteq (x \times y)$ such that $R_1``y = x$ and ${R_1}^{-1}``x = y$. At stage n, player $\neq$ picks a pair $\langle x_n, y_n \rangle$ from $=$'s previous choice; subsequently player $=$ chooses $R_{n+1} \subseteq (x_n \times y_n)$ such that $R_{n+1}``y_n = x_n$ and ${R_{n+1}}^{-1}``x_n = y_n$. Player $=$ loses if she is confronted with a pair $\langle x_n, y_n \rangle$, one of which is empty and the other not or which are distinct *urelemente*; $\neq$ loses if he picks $\langle x_n, y_n \rangle$, both of which are empty. If the game goes on for ever then $=$ wins.

$G_{x=y}$ is an open game. That is to say, if player $\neq$ wins at all, he has done so after finitely many moves. So $\neq$ or $=$ must have a winning strategy. It is not hard to see that the relation

$$= \text{ Wins } G_{x=y}$$

is a contraction and, according to the minimal rubbish assumption, must be equality.

This gives us the

Axiom of strong extensionality

$$(\forall x)(\forall y)(x = y \longleftrightarrow = \textit{ Wins } G_{x=y}).$$

The axiom of strong extensionality is a very powerful way of getting rid of sets which, like Quine atoms, are ill-founded in apparently gratuitous ways. Let us say x is *bad* if its transitive closure does not contain Λ (in these circumstances there is clearly only one empty set). Obviously all members of bad sets are bad. Therefore, if x and y are both bad then $=$ Wins $G_{x=y}$ by always playing $x_n \times y_n$. So all bad sets are equal, and the unique bad set must be a Quine atom. But we have already been told by pseudoinduction that there are no Quine atoms, so every transitive set contains Λ.

There are natural connections between these games.

REMARK 1.2.7 *If* $=$ *Wins* $G_{x=y}$ *then whichever of* I *and* II *Wins* G_x *also Wins* G_y.

Proof: Suppose $=$ Wins $G_{x=y}$. I, say, Wins G_x. To Win G_y he makes a winning move in G_x, and uses $=$'s winning strategy in $G_{x=y}$ to get a move in G_y. There may be several, of course. II's reply is copied back via $=$'s strategy. I replies in G_x and so on. I eventually wins the play in G_x so I's last choice in G_x is an empty set. Now the maps used have been fragments of Winning strategies for $=$, and such maps pair empty sets with empty sets; so I's last move in G_y was to an empty set too, and so he has won G_y. ■

REMARK 1.2.8 *If x and y are in* I ∪ II *then* $=$ *Wins* $G_{x=y}$ *iff* $x = y$.

Proof: One direction is easy and we prove the other by induction on pseudorank. ■

Another natural principle that one could call an axiom of pseudofoundation is the assertion that, for any X, the collection of things that are hereditarily in X is either V or included in *WF*. We shall see later how this has no definable counterexamples in *NF* (see proposition 2.1.8).

In contrast, Hinnion ([1981] onwards) and Forti and Honsell (see the bibliography) consider axioms whose effect is that the class of contractions is rich. For example, every extensional relation is the $\in$-graph of a transitive set. I should conclude this section by pointing out that these axioms of pseudofoundation and strong extensionality are not routinely part of any of the set theories with $V \in V$ to which this volume is devoted and, although their consistency will occasionally be discussed, they will not be adopted.

1.2.2 *Sets as natural kinds*

The root of *topological set theory* (see the writings of Forti, Hinnion, and Weydert in the bibliography) is the idea that sets can be seen as classes which have nice closure properties. One obvious defect with the sets-as-predicates allegory, at least in the form in which I took it earlier, is that we are interested specifically in mimicking the attribution not of *arbitrary* predicates, but of what have recently come to be called "natural kind" predicates. There is an extensive philosophical literature on natural kinds, and this device has been called upon to do a lot of work. For example, the difference between a law of nature and other (ordinary) $\forall_1$ sentences is supposed to be that the predicates in the law of nature are natural kind predicates. Since the complement of a natural kind is not necessarily a natural kind, we can discern a difference between a black raven helping to confirm that all ravens are black (to use a notorious example), whereas the arrival of a white thing that turns out to be a butterfly does not help to confirm that all non-black things are not ravens, since non-black is presumably not a natural kind.

Anyone with a mathematical training who looks at this gadgetry is immediately going to think that in a formalized version of the notion of natural kind, the kinds will be characterized topologically, perhaps as regular open sets. Consideration of the standard puzzles will tell us quite a lot about how this topology should behave. *Via* the idea that a "good" piece of default reasoning can be characterized by one with a topologically small set of counterexamples, principles of default reasoning (should anybody ever find any) could tell us a lot about this topology.

Pleasing though this idea is, it is very difficult to assimilate the difference between natural kind and arbitrary predicate to the difference between a set and a class, for the simple but silly reason that set-theoretically the difference between a set and a class is that proper classes are not members of anything and therefore we do not predicate things of them. This is not the same difference at all. It is true that using the natural kind/arbitrary predicate distinction as a source of axioms for set theory could indeed give rise to a two-sorted theory of extensional objects formally like that of sets and classes, but the classes (the second sort) would not be identifiable as sets-not-belonging-to-anything as they are in traditional theory of sets and classes. The result would bear very little similarity to set theory with classes of the kind we have known so far.

None of these reflections seem to be what is causing Forti, Hinnion, and Weydert to go down this path. It is pleasing to note, though, that their notion of topological set theory does reproduce one feature that we have already identified as desirable in a theory of sets-as-natural-kinds that is not desirable in a theory of sets-as-predicates, namely that, although V

is a set, in general the complement of a set is not a set. What is more, this feature arises in the same way in both cases, namely that for both developments is it necessary that the topology not be discrete.

1.3 A brief survey

The first formal system of set theory with a universal set (if we except Frege!) is *NF* in Quine [1937a], from whose title ("New foundations for mathematical logic") it takes its name. For philosophically motivated readers this article is still worth reading (not recommended otherwise, for it is technically misleading). Quine has written recently (Quine [1987]) about how he arrived at the idea of *NF*. This is one of those cases where the simplest pedagogical approach to a topic happens also to be the historically correct one. The simplest way to present *NF* is to suggest that, if we restrict ourselves to instances of the comprehension scheme

$$\forall \vec{x} \exists y \forall z (z \in y \longleftrightarrow \phi(\vec{x}, z))$$

where ϕ satisfies a typing discipline borrowed from Russell (to wit, ϕ is stratified), then that is enough to avoid the paradoxes even if we make no other use of the typing discipline, so that the theory is one-sorted. Given this, the obvious first question is: "What becomes of Russell's paradox?", and the obvious answer is that since '$x \notin x$' is not stratified, $\{x : x \notin x\}$ is not automatically a set. First hurdle cleared.[14]

As Quine explains it, his starting point was the theory referred to in this book as TST. An obviously unattractive feature of this system is the reduplication of entities (V, Λ, etc.) at each type. Interestingly he suggests that it may have been from Zermelo that he got the idea that one could discriminate between the two properties of formulae: (i) that of being well-formed and (ii) that of determining a set. Once this distinction has been made, it becomes possible to use the typing discipline merely to restrict the comprehension scheme and not restrict the well-formation rules at all. Even now, 57 years later, Quine's intuition that this would be sufficient to avoid the paradoxes has still not been proved wrong.

In particular, there is no specific propaganda campaign arguing that there is a philosophical justification for the axioms of *NF*. Although *NF* has a finite axiomatization saying that V is closed under certain operations, there is no feeling that sets are being constructed in accordance with them (in however aetiolated a sense of 'constructed'), in contrast with *ZF* and, as we shall see later, Church's set theory *CUS*.

Specker [1953] discovered that *NF* refutes the axiom of choice. This is odd because the type theory of Russell from which Quine borrowed

[14] Of course it might turn out to be a set for other reasons! That possibility remains to be excluded.

the typing is consistent with *AC*. Specker also noticed that *NF* (which is a one-sorted theory) is equiconsistent with Russellian type theory with an additional axiom scheme saying roughly that all types are elementarily equivalent. Specker's equiconsistency theorem is the foundation for all study of the models of *NF*, and his refutation of *AC* in *NF* is the source of all worry that there might not be any models to study. Various natural subsystems of *NF* have been proposed and investigated but, at the time of writing, the only provably consistent fragment of *NF* in which any of the known pathologies of *NF* can be reproduced (such as the proof of the axiom of infinity which can be proved without deriving it as a corollary of $\neg AC$) is in Crabbé [1982a].

Since 1974, three non-trivial set theories with a universal set have appeared which arise in a different tradition. These are Church's *CUS* [1974] and the others inspired by it: Mitchell [1976] and the system of Sheridan's Oxford D.Phil. thesis [199?]. These theories all arise out of a technique of defining a new $\in$-relationship over a model of *ZF* or in principle any theory with the axiom of foundation. The idea is to have a bijection f between V and some complicated class of n-tuples, as it might be $V^3 \times On \times [\mathbb{N}]^{\leq\omega}$. Next define x "in" y iff x stands in some relation to the components of $f`y$, where this relation is to be spelled out in detail in each case. Not only have the theoretical limits of this technique not been ascertained, but only a very few examples of what can be done with this have even been looked at. Still, a picture emerges. It is relatively easy to devise f so that in the resulting model the well-founded sets are a copy of the original model, or even that the axiom scheme of replacement holds for well-founded sets. It is easy to verify pairing, for example. This is because such things are often inherited free from the original model. If we wish to arrange for given constructions to be possible for big sets we have to do this by cleverly jigging the use of the components of the n-tuple. This is hard work, and it is sad to have to report that, for all the models constructed so far in this spirit, the corresponding theories are pretty trivial once we delete from them all axioms making explicit reference to well-founded sets. However, the intimate relation that (models of) these theories bear to (models of) *ZF* means that much of the philosophical propaganda about the cumulative hierarchy that we all know from *ZF* can be recast with little effort. In particular, it is still possible to think of the universe as being constructed by transfinite recursion from a handful of operations. Basically this happens because the kind of theory of big sets that these constructions can support is so impoverished that the equational theory of the free algebra over these operations is very simple, and it is correspondingly easy to discover when two n-tuples code the same set.

Another area was first explored by J. Malitz, and more recently treated thoroughly by Weydert (see the bibliography). Define over *On* a family of equivalence relations on V by

$$\sim_0 = V \times V$$
$$\sim_{\alpha+1} = (\sim_\alpha)^+$$

taking intersections at limit ordinals. See definition 1.2.6 for explanation of the "+" operation. For α suitably large we take a quotient. It turns out that this is a model of extensionality and $V \in V$, and satisfies various axioms in the tradition of topological set theory (see Forti and Hinnion [1989] and Weydert [1989]).

1.4 How do theories with $V \in V$ avoid the paradoxes?

Plainly the existence of a universal set is incompatible with an unrestricted *separation* scheme. (The use of the word 'separation' is preferable here to "comprehension" because of the latter word's use in denoting schemes like $\forall\vec{x}\exists y\forall z(z \in y \longleftrightarrow \Phi(\vec{x}, z))$. *Separation* is the scheme: $\forall\vec{x}\forall w\exists y\forall z(z \in y \longleftrightarrow (z \in w \wedge \Phi(\vec{x}, z)))$.

Mirimanoff's paradox, the paradox of the class of all well-founded sets, is resolved in the same way by all the set theories under consideration here: the class of well-founded sets is not a set.

Russell's paradox depends on the existence of $\{x : x \notin x\}$. None of the set theories here have such an axiom. *NF* avoids it because it is unstratified, positive set theory because it is not positive.

Cantor's paradox is more interesting. If we can prove Cantor's theorem, that $\mathcal{P}`x$ is strictly bigger than $x$, we have a contradiction with the existence of $V$, since there is no way whatever of escaping the fact that $V = \mathcal{P}`V$. Cantor's theorem depends on a diagonal argument. We suppose $f : x \to \mathcal{P}`x$ surjectively, and consider $\{y \in x : y \notin f`y\}$. If this is a set, then the contradiction follows by elementary logic. What must happen if we are to escape paradox is that this instance of comprehension fails to be a theorem. In *NF* (resp. positive set theory) this is because the formula is unstratified (resp. not positive); with theories of the Church–Mitchell variety it is essentially because they have no comprehension scheme whatever for large sets, though they do have a version of the axiom scheme of replacement for sets the same size as well-founded sets.

The Burali-Forti paradox is the only classical paradox which our systems evade by radically different means. Church–Mitchell and positive set theory do not allow that *On* is a set, so we have no problem. In this respect, as in so many others, they behave like *ZF*. *On is* a set in *NF* (and we use a different notation '*NO*' to remind ourselves of this). It therefore has an ordinal number. The escape route is that in *NF* there seems to be no way of showing that each ordinal counts the sequence of its predecessors in their natural ordering. Without this there is, so far, no paradox.

We remarked earlier that we have to take very seriously the treatment of truth definitions in set theories with $V \in V$. Set theory is a one-sorted theory in which arithmetic can be coded. It therefore prima facie offers us the possibility of proving a formalized version of the liar paradox. It is always an instructive exercise to attempt this, on meeting a new set theory T. Of course, if T is consistent, then this will fail, but the manner of the failure can tell us much about the system, for the bits of exploded proof will usually spontaneously reassemble into theorems important in their own right. In *ZF* we get the various reflection theorems and the Levy hierarchy theorem; in *NF* we get the theorem (theorem 2.3.8) of Orey on the consistency of related fragments of type theory.

1.5 Chronology

[1937] Quine publishes "New foundations for mathematical logic".
[1944] Hailperin finds a finite axiomatization for *NF*.
[1953] Specker proves $\neg$AC in *NF*.
[1960] Specker reduces the consistency problem for *NF* to that of type theory plus full ambiguity. First application of Rieger–Bernays permutation methods to *NF* by Scott.
[1964] Orey's application of truth definitions to *NF* proves the independence of the axiom of counting.
[1967] Henson's Ph.D. thesis on *NF*.
[1969] Jensen's consistency proof for *NFU* and Grishin's consistency proof for NF_3 inaugurate the study of subsystems of *NF*.
[1970s] Formation of the Séminaire *NF* under the leadership of Boffa in Brussels. Work on proof theory by Crabbé. Ph.D.s by Oswald, Forster, Pétry, Hinnion.
[1974] Church's interpretation of his set theory with $V \in V$ in ZFC. Skala's set theory.
[1976] Mitchell's set theory. Malitz set theory.
[1980s] Combinatorial explosion. Topological set theory.
[1987] First meeting devoted specifically to *NF*, Oberwolfach.
[1990] Holmes' Ph.D. thesis on systems of combinatory logic related to *NF*.
[1993] Dzierzgowski's Ph.D. thesis on intuitionistic systems related to *NF*.
[1994] Jamieson's Ph.D. thesis.

2

NF AND RELATED SYSTEMS

2.1 *NF*

The axioms of *NF* are extensionality plus all stratified instances of the scheme

$$\forall \vec{x} \exists y \forall z (z \in y \longleftrightarrow \Phi(\vec{x}, z))$$

with y not free in Φ. What does this actually give us? Well, V is a set, and it is a complete boolean algebra under set inclusion. That is to say, if x and y are sets, so are $x \cup y$, $x \cap y$, $-x$, $-y$, $\bigcup x$, $\bigcup y$, $\bigcap x$, and $\bigcap y$. Also $\iota`x$ is a set, $\mathcal{P}`x$ is a set, and dually the set $F`x$ of supersets of any set x is a set. It is helpful to think of these as forming a circus stage on which we can perform the feats of set existence licensed by the scheme of stratified comprehension. The stage at least is known to be sound (all the axioms mentioned are in NF_3 of which more in section 2.3.1) even though the status of the plot is problematical. As is usual in these matters, various people have at various times believed they have proved the consistency of *NF* or believed they have proved its inconsistency. Kuzichev and Schultz (who both thought they had proved the consistency of *NF*) got as far as producing material that can be cited. See Kuzichev [1983] and references therein, and Schultz [1980]. Nobody seems to believe any of it.

NF is finitely axiomatizable, though not much is made of this fact. The following axiomatization is due to Hailperin [1944]:
P1: $\forall u \forall v \exists y \forall x (x \in y \longleftrightarrow (x \notin u \vee x \notin v))$
P2: $\forall u \exists v \forall x \forall y (\langle \iota`x, \iota`y \rangle \in v \longleftrightarrow \langle x, y \rangle \in u)$
P3: $\forall u \exists v \forall x \forall y \forall z (\langle x, y, z \rangle \in v \longleftrightarrow \langle x, y \rangle \in u)$
P4: $\forall u \exists v \forall x \forall y \forall z (\langle x, z, y \rangle \in v \longleftrightarrow \langle x, y \rangle \in u)$
P5: $\forall u \exists v \forall x \forall y (\langle x, y \rangle \in v \longleftrightarrow x \in u)$
P6: $\forall u \exists v \forall x (x \in v \longleftrightarrow \forall z (\langle z, \iota`x \rangle \in u))$
P7: $\forall u \exists v \forall x \forall y (\langle y, x \rangle \in u \longleftrightarrow \langle x, y \rangle \in v)$
P8: $\exists v \forall x (x \in v \longleftrightarrow \exists y (x = \iota`y))$
P9: $\exists v \forall x \forall y (\langle \iota`x, y \rangle \in v \longleftrightarrow x \in y)$.

One rather unattractive feature of this axiomatization is that it depends crucially on ordered pairs being Wiener–Kuratowski, that is $\langle x, y \rangle = \{\{x\}, \{x, y\}\}$. Since it does not seem to have attracted much attention, and the original is accessible and readable, we will not go over it here. Beneš [1954] has a consistency proof of P1–P8. P6 is the one very problematic axiom, and it is not easy to see what it is doing. Lake has a consistency proof of P1–P5 and P7–P9 which we shall allude to later (section 2.3.2). Crabbé has a finite axiomatization of *NF* which is less sensitive to the use of Wiener–Kuratowski ordered pairs, but he has not published it. Another finite axiomatization arises from theorem 2.3.26.

The first question in the mind of newcomers to *NF* concerns the paradoxes. We have seen in section 1.4 how *NF* avoids Russell's paradox. Our first task here is to show that a version of Cantor's theorem can nevertheless be proved: all sets have fewer singleton subsets than subsets.

PROPOSITION 2.1.1 *NF* $\vdash \forall x$ *there is no surjection from* $\iota``x$ *onto* $\mathcal{P}`x$.

Proof: It is clear from the axioms of *NF* that if x is a set so is $\iota``x$. Let us use Wiener–Kuratowski ordered pairs until further notice and suppose we have a surjective map $f : \iota``x \to \mathcal{P}`x$. Consider the expression "$\{y : y \in x \land y \notin f`\iota`y\}$". By writing this out in primitive notation the reader can check that "$y \in x \land y \notin f`\iota`y$" is stratified with $f$ being assigned a type $n+4$ when $y$ has type $n$. '$\{y : y \in x \land y \notin f`\iota`y\}$' therefore denotes an object, and one which by the usual diagonal argument cannot be in the range of f. ■

As remarked in section 1.4, if we try to prove the usual version of Cantor's theorem the diagonal set has an unstratified defining property.

COROLLARY 2.1.2 $\iota``V$ *is strictly smaller than* V.

Proof: V is its own power set. ■

Cardinals in *NF* will be equivalence classes under equinumerosity: the cardinal of x, $\overline{\overline{x}}$, is the set of all things the same size as x, and it is easy to check that the existence of $\overline{\overline{x}}$ is guaranteed by a stratified set existence axiom. We write '$\overline{\overline{x}} \leq \overline{\overline{y}}$' iff there is a 1-1 map from x into y and '$\overline{\overline{x}} \leq^* \overline{\overline{y}}$' iff there is a many–one map from y onto x or x is empty.

We shall see later that $\overline{\overline{\iota``x}} \leq \overline{\overline{x}}$ is not provable in general, for it implies Con(*NF*). Sets $x$ such that $\iota``x$ and x *are* the same size are said to be *cantorian* and a large part of understanding the running of *NF* is simply understanding which sets are cantorian. We write "$can(x)$" for short. A set is *strongly cantorian* iff the restriction of ι to $x \times \iota``x$ is a set, that is to say the restriction of the singleton function is a set and witnesses $can(x)$. We write "$stcan(x)$" similarly. We shall see later that, although we can prove in *NF* that there are infinite cantorian sets ($\mathbb{N}$ is one), if *NF* is consistent we cannot prove in *NF* that there are infinite strongly cantorian sets nor that there is an infinite well-founded set. (see proposition 3.1.16).

The following are left as exercises:

1. If $can(x)$ and there is a bijection between x and y then $can(y)$.

2. $can(x) \rightarrow can(\mathcal{P}`x)$.
3. $can(x) \wedge can(y) \rightarrow can(x \times y)$.
4. $can(x) \wedge can(y) \wedge (x \cap y = \Lambda) \rightarrow can(x \cup y)$;

but not $can(x) \wedge can(y) \rightarrow can(x \cup y)$ (which appears to be open) or $can(x) \wedge can(y) \rightarrow can(x \cap y)$! Since it is easy to show that a pair of countable sets can have intersection any finite size, this last would imply that all finite sets are cantorian. We also have analogous additional closure properties of strongly cantorian sets.

5. If $stcan(x)$ and there is a bijection between x and y, then $stcan(y)$.
6. $stcan(x) \wedge \overline{\overline{y}} \leq \overline{\overline{x}} \rightarrow stcan(y)$.
7. $stcan(x) \wedge stcan(y) \rightarrow stcan(x \cup y)$.
8. Any surjective image of a strongly cantorian set is strongly cantorian.

A natural piece of exploration in any new set theory is to see how one has to do arithmetic. Most of the definitions we will need have already been presented in the introduction. One that has not (for not all set theories with a universal set do arithmetic this way) is the definition of the Russell–Whitehead integers

$$0 =_{\text{df}} \{x : \forall y(y \notin x)\}$$

and

$$S`x =_{\text{df}} \{z : (\exists w \in z)((z - \iota`w) \in x)\}.$$

The reader may verify that these definitions are stratified, and that their effect is that 0 is the set of all empty sets (i.e. $\iota`\Lambda$), 1 ($S`0$) is $\iota``V$, the set of all singletons, and generally n is the set of all n-membered sets. We may then define

$$\mathbb{N} = \bigcap\{x : 0 \in x \wedge S``x \subseteq x\}.$$

We shall see in due course that all of finite-order arithmetic can be successfully interpreted into *NF*, but this needs the proof of the axiom of infinity which will require some work.

Now we can define Quine ordered pairs. Let $\theta_1`x =_{\text{df}} (x - \mathbb{N}) \cup S``(x \cap \mathbb{N})$ and $\theta_2`x =_{\text{df}} (x - \mathbb{N}) \cup S``(x \cap \mathbb{N}) \cup \iota`0$.

The ordered pair $\langle x, y\rangle$ is defined to be $\theta_1``x \cup \theta_2``y$. The decoding function depends on the fact that nothing can be a value of both θ_1 and θ_2. If $x = \langle y, z\rangle$ then $y = \theta_1^{-1}``(x \cap \theta_1``V)$, and z similarly. The great virtue of this definition is that it is *homogeneous*: '$z = \langle x, y\rangle$' is stratified with all three variables having the same type. It also emerges that, with this definition, everything is an ordered pair; indeed everything is a *set* of ordered pairs. One consequence of this will be that we will eventually be able to prove that there is a 1-1 correspondence between V and V^V, the set of all total maps from the universe into itself. This raises the natural question as to whether or not, under some such map, the universe is a natural model of untyped λ-calculus. Unfortunately the set existence axioms we would apparently need to verify this are unstratified ("ill-typed"), so it is not obviously true. But it is not obviously false either!

However, none of this machinery can be invoked until we have shown that it works. Proving uniqueness will clearly depend on $\mathbb{N}$ having infinitely many distinct members, in short, the axiom of infinity (first noted by Rosser [1952]). This is theorem 2.2.7. We must take care that none of the machinery used in its proof depends on $\langle x, y\rangle$ being a Quine pair, not a Wiener–Kuratowski pair.

We define $\cdot$ and $+$ in the usual way once we have Quine ordered pairs, that is

$$\overline{\overline{(x \times y)}} = \overline{\overline{x}} \cdot \overline{\overline{y}}$$

and

$$\overline{\overline{x}} + \overline{\overline{y}} = \overline{\overline{(x' \cup y')}}$$

where $\overline{\overline{x}} = \overline{\overline{x'}}$, $\overline{\overline{y}} = \overline{\overline{y'}}$, and $y' \cap x' = \Lambda$. Since '$z = \langle x, y\rangle$' is stratified with all three variables having the same type, "$z = x \times y$" is too. This enables us to show (given Quine ordered pairs and the axiom of infinity) that $\langle \mathbb{N}, +, \cdot, 0, S\rangle$ is (provably in *NF*) a model of nth-order arithmetic for every standard n. This is because any expression $\Psi(\vec{x})$ from nth-order arithmetic where the ith variable is of order t_i (i.e. ranging over elements of $\mathcal{P}^{t_i}\text{‘}\mathbb{N}$) is stratified: in any stratification, 'x_i' will be t_i types above any variable over integers. Therefore, by comprehension in *NF*, this structure satisfies the appropriate comprehension axioms of nth-order arithmetic and, by considering the set $\{x \in \mathbb{N} : \mathbb{N} \models \psi(x)\}$ where ψ is an expression in nth-order arithmetic, possibly containing parameters, we see that this structure satisfies the induction axiom.

We can define exponentiation in $\mathbb{N}$ by recursion over $\mathbb{N}$ in the usual way, but there is some interest attached to the question of how to do it for infinite cardinals. It will be helpful to have a definition of "$x = 2^y$" that makes 'x' and 'y' the same type, so that the general definition is compatible with what we obtain for $\mathbb{N}$ by recursion. We will define $x = 2^y \longleftrightarrow_{\mathrm{df}} \exists w(y = \overline{\overline{\iota\text{“}w}} \wedge x = \overline{\overline{\mathcal{P}\text{‘}w}})$. This has the consequence that 2^x is undefined for certain x (e.g. $\overline{\overline{V}}$ and indeed any cardinal $\not\leq \overline{\overline{\iota\text{“}V}}$). The advantage is that we can define lots of things which we would not have been able to do with the total but inhomogeneous version "$x = 2^y \longleftrightarrow_{\mathrm{df}} \exists w(y = \overline{\overline{w}} \wedge x = \overline{\overline{\mathcal{P}\text{‘}w}})$". For example, for a cardinal α we define

$$\mathcal{T}\text{`}\alpha = \bigcap\{y : \alpha \in y \land \forall\beta(2^\beta \in y \to \beta \in y)\}$$

with the associated partial order

$$\bigcap\{R : (\forall zy \in \mathcal{T}\text{`}\alpha)(z = 2^y \to zRy) \land (\forall xyz \in \mathcal{T}\text{`}\alpha)[(xRy \land z = 2^y) \to xRz]\}.$$

In English, we build a tree below a cardinal α recursively by placing immediately below each β in the tree all the γ such that $2^\gamma = \beta$. We define it in this direct way in order not to have to appeal to the axiom of infinity, which we have not yet proved, and for the proof of which this tree structure will be crucial. Note the characteristic use of intersection over all sets closed under a certain operation to obtain an inductively defined set, in contrast to the *ZF* style in which such objects are typically obtained by recursion over naturals, ordinals, or other well-founded structures.

2.1.1 *The axiom of counting*

In set theory proofs of '$\forall x\Phi$' by induction over a well-founded structure M require that the set $\{x \in M : \neg\Phi\}$ of counterexamples be a set, and are therefore sensitive to comprehension axioms. Thus when we turn to induction over the natural numbers, although we have no trouble with $+$, $\times$, etc., since these are coded by stratified functions, we can expect trouble in more general cases since we can perform induction prima facie for stratified formulae only. For example, we must not expect to be able to prove that each $n \in \mathbb{N}$ counts the set of its predecessors: '$\{m : m < n\} \notin n$' is not stratified. We shall see in section 2.3.1 how this particular piece of induction proves the consistency of *NF* and cannot be performed if *NF* is consistent. This raises the intriguing possibility that the integers of *NF* may be non-standard in ways that the theory itself can actually describe. Consider the function T on cardinals, where $T\text{`}\,\overline{\overline{x}} =_{\text{df}} \overline{\overline{\iota\text{``}x}}$. Then $T$ cannot exist as a set of ordered pairs (unless it is $=$) but we can still manipulate formulae containing '$T$'. Evidently $T$ restricted to $\mathbb{N}$ defines an automorphism of[15] $\mathbf{N}$, and $T\text{``}NC$ is going to be an isomorphic copy of NC. There is an analogous function defined on ordinals (or for that matter any kind of relational type), which we will also notate 'T': if α is the order-type of $\langle X, R\rangle$, then $T\text{`}\alpha$ is the order-type of $\langle \iota\text{``}X, RUSC(R)\rangle$ (in Henson [1973a] this is 'U'). The assumption that T is the identity on $\mathbb{N}$ can make life a lot easier. Because it is equivalent to $(\forall n \in \mathbb{N})(\{m : m < n\} \in n)$, Rosser [1953a] called it the *axiom of counting*, or AxCount for short. It

[15]It is obvious that T respects $+$, $\times$, and S. Once we have the axiom of infinity we can see easily that both $T\text{``}\mathbf{N}$ and $T^{-1}\text{``}\mathbf{N}$ contain 0 and are closed under S. Thus $\mathbf{N} \subseteq T\text{``}\mathbf{N} \cap T^{-1}\text{``}\mathbf{N}$ and so $\mathbf{N} = T\text{``}\mathbf{N}$.

is often called 'R' (for "Rosser") by the Belgian school. AxCount does not license *all* inductions over $\mathbb{N}$ (it does not enable us to make "n has a member which is hereditarily finite" look stratified) but it does mean that if the unstratified Φ we are trying to prove by induction is unstratified only because it contains a subformula '$\psi(n,m)$', where n and m are integers and 'n' is k types higher than 'm', then we can still perform the induction. This is because, if we have AxCount, $T^k m = m$, so, by use of substitutivity of equality to replace '$\psi(n,m)$' by '$\psi(n,T^k m)$' we have reduced the problem to a stratified induction. *NFC* is *NF* plus the axiom of counting.

Although the proof of the following theorem uses the axiom of infinity (whose proof is a long way off), we shall look at it now in order to orient ourselves in this strange landscape. However, we must be careful not to use it until we have proved the axiom of infinity.

PROPOSITION 2.1.3 *In NF the following are equivalent:*

1. *All finite sets are strongly cantorian.*
2. *All finite sets are cantorian.*
3. *There is an infinite strongly cantorian set.*

Proof: Recall that a finite set is one with cardinal in $\mathbb{N}$. $1 \to 2$ is obvious. If 2 holds then $Tn = n$ for all naturals, so $\{\langle Tn, \iota\text{'}n\rangle : n \in \mathbb{N}\}$ (which is a set anyway) is simply ι restricted to $\mathbb{N}$ and so $\mathbb{N}$ is a strongly cantorian set. This gives $2 \to 3$. By induction over $\mathbb{N}$ any infinite set has subsets of all finite sizes (if it has subsets of size n but not of size $n+1$ it has precisely n members), and any subset of a strongly cantorian set is likewise strongly cantorian. Therefore every finite set is the same size as a strongly cantorian set and is therefore strongly cantorian. (See exercise 5 on page 27.) Thus 3 implies 1. ■

In a set theory with a universal set, there must be proper classes that are included in sets.[16] It can turn out that these proper classes can be finite. We shall see that AxCount is independent of *NF* (theorem 2.3.8); let us assume this for the moment and work in $NF + (\exists n \in \mathbb{N})(n \neq Tn)$. Fix such an n. If $\{m \leq n : m = Tm\}$ is a set, then we prove by induction on the integers below n that $(\forall m < n)(m = Tm)$, and so $n = Tn$, contradicting assumption $n \neq Tn$. Thus if AxCount fails, there will be finite proper classes. Of course this will happen in any model of a set theory where $\mathbb{N}$ is non-standard and has an external automorphism. We shall see more interesting examples than this later.

There are two weaker versions of AxCount which should be considered, one much more important than the other. $\text{AxCount}_{\leq}$ is the assertion that $(\forall n \in \mathbb{N})(n \leq Tn)$ and $\text{AxCount}_{\geq}$ is the assertion that $(\forall n \in \mathbb{N})(n \geq Tn)$.

[16] Such proper classes are sometimes called *semisets* (see Vopěnka and Hájek [1972], Vopěnka [1979]).

AxCount$_\leq$ seems to be much stronger than AxCount$_\geq$, and is more natural than AxCount, in that it turns out to be equivalent to many apparently disparate assertions. We can prove some of these equivalences here, but the rest must await the terminology and techniques of the section on classes of invariant formulæ (see section 3.1.1.2) where we shall prove theorem 3.1.28. Let us write "$x <^T y$" for "$Tx < y$".

THEOREM 2.1.4 *The following are equivalent:*

1. *AxCount$_\leq$.*
2. *$<^T$ is well-founded on $\mathbb{N}$.*
3. *Classical Cantor's theorem ($\overline{\overline{x}} < \overline{\overline{\mathcal{P}`x}}$) holds for finite sets (sets with cardinal in $\mathbb{N}$).*

Proof: $1 \longleftrightarrow 2$

Assume AxCount$_\leq$ and suppose the relation $Tx < y$ is not well-founded. Then there is a non-empty $X \subseteq \mathbb{N}$ such that $(\forall n \in X)(\exists y \in X)Ty < n$. But X must have a $<$ minimal element n_0 say. Therefore for some $y \in X$, $y \geq n_0$ but $Ty < n_0$. Therefore AxCount$_\leq$ fails. Now let us suppose the relation is well-founded. Consider the singleton $\iota`n$ for some $n \in \mathbb{N}$. This must have a minimal element, so $\neg(Tn < n)$, which is to say $n \leq Tn$ but n was arbitrary.

$1 \longleftrightarrow 3$

Evidently 3 is equivalent to the assertion that $n < 2^{Tn}$ for all $n \in \mathbb{N}$. Now consider finite beth numbers. If $n > Tn$ then $n > Tn{+}1$. ($n = Tn{+}1$ is impossible since the two sides are not congruent mod 2.) Since $\beth$ commutes with T, $\beth_n`0 > \beth_{Tn+1}`0$. Abbreviating '$\beth_n`0$' to 'm', we have $m > 2^{Tm}$. If this is impossible by 3, then $n > Tn$ cannot happen either, so $3 \rightarrow 1$. The other direction is easy. ■

By virtue of 2, AxCount$_\leq$ can be seen as embodying an induction principle. (AxCount$_\geq$ is dually equivalent to the assertion that the relation $n < Tm$ is well-founded (on $\mathbb{N}$). The proof is the same and is omitted.)

AxCount$_\leq$ implies not only that that the relation $<^T$ is well-founded, it implies that it is a well-quasi-order. (We will not make use of this fact.) It is transitive because if $Tn < m \wedge Tm < k$ then $T^2n < k$ and $k \leq Tk$ so $T^2n < Tk$ and $Tn < n$. For the condition concerning ω-sequences let $\langle x_i : i \in \mathbb{N}\rangle$ be an ω-sequence of distinct natural numbers. By AxCount$_\leq$ it has a $<^T$-minimal element, n, say. Let X be the set of elements of $\langle x_i : i \in \mathbb{N}\rangle$ that occur later than n. If there is no $x \in X$ s.t. $n <^T x$ then $(\forall x \in X)(x \leq Tn)$ so X must have been finite.

André Pétry has recently been investigating the relation between the axiom of counting and the scheme $(\forall n \in \mathbb{N})(\phi(n) \longleftrightarrow \phi(Tn))$ over all stratified ϕ and will publish his findings soon. Unlike the axiom of counting this scheme is stratified, and since we know from theorem 2.1 of Henson [1969] (which we treat here in section 3.1.2) that none of AxCount, $\text{AxCount}_{\leq}$, and $\text{AxCount}_{\geq}$ are theorems of consistent stratified extensions of *NF*, the scheme cannot imply AxCount. At this stage, it appears to be an open question whether or not it implies all the stratified consequences of AxCount.

The scheme arises naturally in connection with a proof due to Orey [1964] of part of Henson's result just referred to. We shall see this in section 2.3.1 and mention Pétry's scheme again.

2.1.2 *Boffa's lemma on n-formulae, and the automorphism lemma for set abstracts*

LEMMA 2.1.5 *If Φ is stratified then*

$$\Phi(x_1, \ldots, x_k) \longleftrightarrow \Phi((j^{n_1}`\sigma)`x_1, \ldots, (j^{n_k}`\sigma)`x_k)$$

for any setlike permutation σ, where n_k is the integer assigned to the variable 'x_k' in some fixed stratification.

Proof: The reader can check that, if $M \models NF$ and σ is a setlike permutation of M, then so is $j^n`\sigma$ for all standard n.

By definition of j we have $x \in y$ iff $\tau`x \in (j`\tau)`y$ for any $\tau$. In particular if '$x$' has been assigned type $n$ and '$y$' the type $n+1$, we invoke the case where $\tau$ is $j^n`\sigma$ to get $x \in y \longleftrightarrow (j^n`\sigma)`x \in (j^{n+1}`\sigma)`y$. By substitutivity of the biconditional we do this simultaneously for all atomic subformulae in $\Phi(x_1, \ldots, x_k)$. Variables 'y' that were bound in '$\Phi(x_1 \ldots x_k)$' now have prefixes like '$j^n`\sigma$' in front of them but, since '$\Phi(x_1, \ldots, x_k)$' was stratified, they will be constant for each such variable '$y$'. We then use the fact that $j^n`\sigma$ is a permutation of V so that any formula $Qy \ldots (j^n`\sigma)`y \ldots$ (Q a quantifier) is equivalent to $Qy \ldots y \ldots$. ■

The trick used in the above proof will also be applied extensively in chapter 3.

We have the following corollary.

COROLLARY 2.1.6 The automorphism lemma for set abstracts. *If Φ is a formula with only 'x' free and has a stratification in which 'x' receives type n then, for any permutation σ of V, $\{x : \Phi(x)\} = (j^n`\sigma)``\{x : \Phi(x)\}$. Equivalently if $\Phi(x)$ then $\Phi(y)$ for any y n-equivalent to x.*

This is quite important. Recall that '$\Phi(x)$' is an *n-formula* iff the smallest integer that can be assigned to 'x' in a stratification is n. If '$\Phi(x)$' is an n-formula such that $\{x : \Phi(x)\}$ codes a structure $\langle V, R \rangle$, then everything in J_n is an automorphism of it. This imposes severe constraints on the kinds of structure we can put on the universe by set abstracts, and we shall see later that it means V can have no definable total orders and no definable well-founded extensional relations.

Remark 2.1.7 *Any term model for NF that is a model of NF (if there are any) must be rigid.*

Proof: By corollary 2.1.6 every term is fixed by all automorphisms. ■

Proposition 2.1.8 Boffa [1971]. *Definable transitive sets (other than V) are well-founded and hereditarily finite of standard rank.*

Proof: Let $\Phi(y)$ be an n-formula so that $\bigcup\{y : \Phi(y)\} \subseteq \{y : \Phi(y)\}$. Then

$$\exists x_1 \ldots x_n (y \in x_1 \in x_2 \in \ldots \in x_n \ \wedge \Phi(x_n))$$

is a 0-formula whose extension is $\bigcup^n \{y : \Phi(y)\}$, which must be transitive if $\{y : \Phi(y)\}$ is. A simple application of lemma 2.1.5 shows that all 0-formulae with one free variable 'y' are equivalent to '$y = y$' or to '$y \neq y$', so these are the only two 0-formulae corresponding to transitive sets. Therefore $\bigcup^n \{y : \Phi(y)\}$ must be V or Λ. If it is Λ, then $\{y : \Phi(y)\}$ is well-founded and hereditarily finite of standard rank; if it is V, then $\{y : \Phi(y)\}$ must be V too. ■

What this shows is that, in term models of *NF* at least, a very strong form of the second axiom of pseudofoundation from page 20 holds: every H_X is either V or included in the union of the V_n (where n is standard) and not merely in *WF* (see section 2.1.4 for the definition of *WF* in *NF*). A similar argument will show that in *NF* there is no set abstract '$\{x : \Phi\}$' whose extension is an infinite well-founded set. For if there were, and Φ is an n-formula, then '$\bigcup \ldots \bigcup \{x : \phi\}$' (with n $\bigcup$) would be another infinite well-founded set, but then it would have to be V or Λ. We shall see later that with more work we can show that it is consistent with respect to *NF* that there should be a (non-standard) finite bound on the size of all well-founded sets.

Proposition 2.1.9 *There are no set abstracts corresponding to $\in$-automorphisms.*

Proof: We can actually prove something stronger than this. Recall the definition of j as $\lambda f.\lambda x.f``x$. A little manipulation will reveal the important fact that a permutation is an $\in$-automorphism iff it is a fixed point for j. We can show that no set abstract can be proved to denote a permutation that can be proved to be, for each standard natural number n, j^n of something, unless it defines the identity.

Suppose that there were such a set abstract: $\{\langle x, y\rangle : \phi(x, y)\}$, where ϕ is a homogeneous formula with 'x' and 'y' of type n in the canonical stratification. Evidently if $\phi(x, y)$ is such a formula then $\psi(x, y)$ defined by $\psi(x, y) \longleftrightarrow \phi(\iota`x, \iota`y)$ is another one, and the automorphism defined by ϕ is j of the automorphism defined by ψ. (If we know that some permutation

σ is j of something τ, to find τ it suffices to see what σ does to singletons.) Clearly the free variables in ψ are one type lower than the free variables in ϕ, so eventually by iteration we will arrive at a definable permutation where the free variables are of type 0, and by lemma 2.1.5 this must be the identity. ■

None of H_{wo}, H_{fin}, etc., are stratified set abstracts because they would be infinite transitive sets distinct from V if sets at all (use proposition 2.1.8).

2.1.3 *Miscellaneous combinatorics*

We will frequently need the following combinatorial fact:

PROPOSITION 2.1.10 *If* $\overline{\overline{X}}=\overline{\overline{Y}}$, *and* $\overline{\overline{-X}}=\overline{\overline{-Y}}$, *then there is a permutation of* V *mapping* X *onto* Y.

Proof: Simply take the union of the two bijections considered as sets of ordered pairs. They are disjoint, total, and onto. ■

The most arresting miscellaneous combinatorial fact in *NF* is the following observation of Boffa's (which is why I have always written the operation with a 'B') which has the status not so much of folklore as of legend.

REMARK 2.1.11 (Boffa)

$$\forall xy(x \in y \longleftrightarrow B`x \in B`y).$$

Proof: $x \in y$ iff (by definition) $y \in B`x$. Repeat the trick to get $y \in B`x$ iff $B`x \in B`y$. ■

This has the immediate consequence that *NF* has no $\subseteq$-minimal model for if $M \models NF$ then $B``M$ is also a model of *NF* (though not a transitive one!) and is a proper substructure of M. The same holds for the operation $\lambda x.\{y : y \notin x\}$. Thus the collection of principal ultrafilters on V is extensional. What about the collection of all ultrafilters on V? Is that a model of *NF*? Is it even extensional?

REMARK 2.1.12

$$\forall x \exists y(\overline{\overline{x}} = \overline{\overline{y}} \ \wedge y \cap \mathcal{P}`y = \Lambda).$$

Proof: If $a \notin a$ then $B`a$ is disjoint from its power set and therefore any subset of it is too. Since it is of size $\overline{\overline{V}}$ it has subsets of all sizes. ■

In the following table we review what is known about some existence questions that have been asked from time to time. Mostly the motivation is immediately evident: well-founded extensional relations are obviously connected to finding natural-looking models of well-founded set theories inside *NF* (see section 2.5); if we want these models to be transitive, we will be interested in x such that $\bigcup x \subseteq x \neq V$. It is a simple exercise to show that in the presence of the axiom of foundation we can prove that there are no non-trivial $\in$-automorphisms of the universe, so it is natural enough to ask what happens once we remove this restriction. Questions about well-quasi-orders and connected antisymmetric relations really reflect our concern about the status of the axiom of choice, for there is a (homogeneously) definable choice function on the set of all pairs iff there is a stratified formula 'Φ' with two free variables such that

$$(\forall xy)(\Phi(x,y) \longleftrightarrow \neg\Phi(y,x))$$

(Φ says: "given x and y, choose y").

THEOREM 2.1.13

	Definable by inhomogeneous stratified formulae	*Definable by unstratified formulae*	*Arbitrary*
Non-trivial automorphisms of $\langle V, \in\rangle$	*No*	*?*	$\langle V, \in\rangle \models \mathrm{AC}_2$ $\rightarrow$ *consistent*
Well-founded extensional relations on V	*Impossible if non-trivial*	*there are automorphisms of* $\langle V, \in\rangle$	*?*
Connected antisymmetric relations on V	*Independent*	*Consistent if NF has a term model*	*?*
Counterexample to pseudo-induction	*Not applicable*	$\{x : x = \iota`x\}$ *can be a singleton*	*consistent*
Infinite well-founded $\bigcup x \subseteq x \neq V$	*Not applicable*	*consistent w.r.t. NFC*	*consistent*

No entities of the above kinds can be definable by homogeneous formulae.

Proof:

(i) Automorphisms

We have shown in section 2.1.2 that there can be no set abstract $\{\langle x,y\rangle : \phi(x,y)\}$ that defines an automorphism of V. The other case—where '$\phi(x,y)$' is stratified but inhomogeneous so that 'x' and 'y' are assigned different types—is not covered by this, since then $\{\langle x,y\rangle : \phi(x,y)\}$

is a class not a set. However, any such Φ would be an inhomogeneous permutation of the universe so that (if 'y' were one type higher than 'x' say) $\{\langle \iota`x, y\rangle : \phi(x,y)\}$ would define a 1-1 map from V onto $\iota``V$, contradicting Cantor's theorem.

(ii) Well-founded extensional relations

No homogeneous $\phi(x,y)$ can define a well-founded extensional relation on V, because no definable total relation is rigid (lemma 2.1.5). Further, if there is an automorphism of $\langle V, \in\rangle$, there can be no definable (even by unstratified means) well-founded extensional relation on V. If σ is an automorphism of $\langle V, \in\rangle$, then σ will fix any definable well-founded extensional relation on V and, if R is a definable well-founded extensional relation on V fixed by σ, we prove by induction on R that σ is $=$. Later (section 2.3.4) we shall show how a certain reduction of the consistency problem for *NF* leads us to consider models of *NF* containing well-founded extensional relations on V.

(iii) Connected antisymmetric relations on V (we will also consider here partial orders where all antichains are of length $< n$)

The treatment of these relations divides into two depending on whether the expression ϕ which is supposed to be defining them is homogeneous or merely stratified. The table warns us that, if ϕ is not stratified, there is little that can be said. First the cases where ϕ is homogeneous.

It is easy to show that, if $\phi(\;)$ is *homogeneous*, then we can *prove* $(\exists x)(\exists y)(\phi(x,y) \longleftrightarrow \phi(y,x))$. The proof given here depends on lemma 2.1.5, the automorphism lemma for set abstracts (which says that every definable structure on V has many automorphisms), though in section 3.1.1 we shall see another (and earlier) proof due to André Pétry. Suppose $\phi(\;,\;)$ is homogeneous with both free variables of type k. Let $\sigma \in J_k$ have a 2-cycle (a,b). There will be such a cycle because j^k of the transposition (u,v) swaps $\iota^k`u$ and $\iota^k`v$. σ is an automorphism of $\phi(\;,\;)$. Suppose $(\forall x)(\forall y)(\phi(x,y) \longleftrightarrow \neg\phi(y,x))$. Then $\phi(a,b) \longleftrightarrow \neg\phi(b,a)$, contradicting the fact that σ is an automorphism of $\phi(\;)$.

To show that there are no homogeneously definable partial orders where every antichain is of length at most n, it will be sufficient to show that there are no such *strict* partial orders. Let ϕ be a homogeneous expression defining a strict partial order where every antichain is of length at most n. Let $\sigma \in J_k$ have a cycle $\langle(\sigma^m)`x : m < p\rangle$ of length some prime $p > n$. This cannot be an antichain, so $\phi((\sigma^m)`x, (\sigma^k)`x)$ for some $m$ and $k$. Therefore also $\phi((\sigma^k)`x, (\sigma^{2k-m (\text{mod } m)})`x)$ etc. But $p$ and $k-m$ are coprime, so eventually $\phi((\sigma^m)`x, (\sigma^m)`x)$, contradicting the fact that ϕ is a strict partial order.

Now for the cases where ϕ need not be homogeneous, but is stratified.

Pétry has shown that, if $\phi(\ )$ is a stratified formula, then we can consistently add to *NF* all axioms

$$(\exists x)(\exists y)(\phi(x,y) \longleftrightarrow \phi(y,x))$$

because any such axiom is a consequence of the assertion that there are at least two Quine atoms, which, as we shall see in section 3.1.1 (proposition 3.1.6), is known to be consistent with respect to any stratified extension of *NF* (at least). Let '$\phi(\ )$' be a stratified formula which defines something connected and antisymmetrical, let x and y be two Quine atoms, and let τ be the transposition (x, y). Then if $\phi(\ )$ defines something connected, we must have either $\phi(x,y)$ or $\phi(y,x)$. Without loss of generality suppose $\phi(x,y)$. Then for some m and n we have $\phi((j^{m}`\tau)`x, (j^{n}`\tau)`y)$, but $(j^{n}`\tau)`y = x$ and $(j^{m}`\tau)`x = y$, so ϕ is not antisymmetric. We shall see in section 3.1.1 how the earlier result that 'Φ' homogeneous implies that '$(\exists x)(\exists y)(\phi(x,y) \longleftrightarrow \phi(y,x))$' is *provable* follows from the fact if 'Φ' is homogeneous then the last quoted expression is stratified under these circumstances.

COROLLARY 2.1.14 *There is no definable choice function on the set of all pairs.*

The non-existence of partial orders where every antichain is of length at most n defined by stratified (but not homogeneous) expressions is proved consistent modulo any stratified extension of *NF* in the same way. As before, it will be sufficient to show that there are no such *strict* partial orders. Let $\phi(x,y)$ be an expression defining such a partial order where 'x' is of type i and 'y' is of type m. Let us assume that there are p Quine atoms $a_1, \ldots, a_p$ for some prime $p > n$. This is consistent by Pétry's result in the same way as before. Consider the cycle $\sigma = (a_0 \ldots a_n)$. In fact, for each k the action of $j^{k}`\sigma$ on $\{a_0 \ldots a_n\}$ is the same as $\sigma$ itself. Since $n < p$ we must have $\phi(a_u, a_v)$ for some $u, v \leq k$. Therefore $\phi((j^{i}`\sigma)`a_u, (j^{m}`\sigma)`a_v)$, which is to say $\phi(a_{(u+1)}, a_{(v+1)})$. By iteration we will conclude that $\phi(a_u, a_v)$ for all u and v, contradicting antisymmetry of $\phi(\ ,\ )$ and distinctness of the a_u.

In any term model of *NF*, we can enumerate the terms (which is to say all sets) by looking at the Gödel numbers of the set abstracts they represent. Thus there will be a (highly unstratified) expression $\Psi(x,y)$ which "represents" a well-ordering of V of order-type ω.[17] Since it is unstratified we are not enabled to define a bijection $V \longleftrightarrow \mathbb{N}$ by recursion on it, so there is no conflict with Cantor's theorem! (See Wang [1953].)

[17]In the sense that we will have $\forall x \forall y(\Psi(x,y) \vee \Psi(y,x))$ and suchlike.

While we are on the subject of these highly asymmetrical objects, it is worth pointing out that we do not know the status of non-principal ultrafilters on V in *NF*. An entry in theorem 2.1.13 would be a row of question marks.

(iv) Counterexamples to pseudoinduction

If X is symmetric then G_X is determinate.

PROPOSITION 2.1.15 *Let X be n-symmetric, with n even (odd). Then either* I (II) *Wins G_X in $n+2$ moves or* II (I) *Wins G_X in $n+3$ moves.*

Let us take the case $n = 6$ as a typical illustration. Let '$\Phi(y, x)$' be short for

$$(\exists x_5 \in x)(\forall x_4 \in x_5)(\exists x_3 \in x_4)(\forall x_2 \in x_3)(\exists x_1 \in x_2)(x_1 \subseteq y).$$

Since II Wins G_x for any $x \subseteq B`\Lambda$, $\Phi(B`\Lambda, X)$ will certainly imply that I Wins G_X (in eight moves in fact). '$\Phi(y, x)$' is a stratified wff in which 'x' is of type 6 and 'y' of type 1. By an application of lemma 2.1.5 (the automorphism lemma for definable sets) we have

$$\Phi(B`\Lambda, X) \longleftrightarrow \Phi((j`\pi)`(B`\Lambda), (j^6`\pi)`X)$$

for any permutation π. But X is by hypothesis 6-symmetric, which is to say $X = (j^6`\pi)`X$ for any π, so this becomes

$$\Phi(B`\Lambda, X) \longleftrightarrow \Phi((j`\pi)`(B`\Lambda), X).$$

Now suppose I does not have a strategy to Win in eight moves. Then $\neg\Phi(B`\Lambda, X)$ and indeed $\neg\Phi((j`\pi)`B`\Lambda, X)$ for any permutation π.

We now seek a permutation π so that $(j`\pi)`(B`\Lambda) = -\mathcal{P}`B`\Lambda$. Since $(B`\Lambda)$ and $-\mathcal{P}`B`\Lambda$ are the same size, as are their complements, there is such a π by proposition 2.1.10. So $\neg\Phi(-\mathcal{P}`B`\Lambda, X)$ which is

$$(\forall x_5 \in x)(\exists x_4 \in x_5)(\forall x_3 \in x_4)(\exists x_2 \in x_3)(\forall x_1 \in x_2)\neg(x_1 \subseteq -\mathcal{P}`B`\Lambda)$$

and the matrix simplifies to $(\exists x_0 \in x_1)(\forall x_{-1} \in x_0)(\Lambda \in x_{-1})$ which is to say II Wins in nine moves. ■

Thus $\in$-determinacy holds in all models of *NF* in which every set is symmetric. In particular, all term models satisfy $\in$-determinacy and if x is (the extension of) a set abstract, G_x is determinate.

(v) Infinite transitive sets $\neq V$

We have already seen (proposition 2.1.8) that all definable transitive sets are V or are hereditarily finite, and we shall see later (theorem 3.1.28) that it is consistent relative to a modest extension of *NF* that V_ω should be a set.

That completes the treatment of the entries in theorem 2.1.13. This next result belongs with them.

PROPOSITION 2.1.16 *The class of all transitive sets is not a set.*

Proof: This proof is due to Boffa, though he has never published it. For some remarks on the author's original proof, see section 2.2.

Suppose *per contra* that Y is the set of all transitive sets. Then we can define $TC`x =_{\text{df}} \bigcap\{y \in Y : x \subseteq y\}$, and $TC$ is a homogeneous function. Let $A = \{x : x \notin TC`(\iota``x)\}$ which is $\{x : x \notin \iota``x \ \wedge \ x \notin TC`x\}$. A has more than one element so $A \notin \iota``A$. If $A \notin TC`A$, then $A \in A$, so $A \in TC`A$. If $A \in TC`A$, then $A \notin A$, so $(\exists x \in A)(A \in TC`x)$ whence $x \in TC`x$ and $x \notin TC`x$ (because $x \in A$). ■

2.1.4 *Well-founded sets*

Well-founded sets have traditionally been characterized by talk about their transitive closures but, since on well-founded sets we can do inductions of the sort $((\forall y \in x\ \Phi(y)) \rightarrow \Phi(x))$, we can equally well characterize *WF* as the intersection of all x such that $\mathcal{P}`x \subseteq x$. Thus

DEFINITION 2.1.17 $WF = \bigcap\{x : \mathcal{P}`x \subseteq x\}$.

This more idiomatic definition is analogous to a definition of $\mathbb{N}$ as the intersection of all sets containing 0 and closed under S. Thus

$$WF(x) \longleftrightarrow_{\text{df}} \forall y(\mathcal{P}`y \subseteq y \rightarrow x \in y).^{18}$$

[18] In the presence of an axiom of complementation we can show that this is equivalent to the usual *ZF* formulation. Thus

$$(\forall y)(\mathcal{P}`y \subseteq y \rightarrow x \in y).$$

Replace y by $-y$ to get

$$(\forall y)(\mathcal{P}` - y \subseteq -y \rightarrow x \notin y)$$

and contrapose

$$(\forall y)(x \in y \rightarrow \mathcal{P}` - y \not\subseteq -y)$$

$$(\forall y)(x \in y \rightarrow y \not\subseteq -\mathcal{P}` - y)$$

$$(\forall y)(x \in y \rightarrow (\exists z)(z \in y \wedge z \in \mathcal{P}` - y))$$

$$(\forall y)(x \in y \rightarrow (\exists z \in y)(z \cap y = \Lambda))$$

but this is of limited interest, since the version we have here is the one needed to justify $\in$-induction.

It is standard to show that if $WF(x)$ then $WF(\bigcup x)$, and that any set of *WF* sets is *WF*. Clearly in the context of *NF* this justifies a scheme of $\in$-induction over *WF* for stratified formulæ. What is interesting is that there are a number of unstratified generalities that one can prove about *WF* (as we shall call the class of all well-founded sets). For example, we shall prove $(\forall x \in WF)(x \notin x)$, because $\mathcal{P}`(-\iota`x) \subseteq -\iota`x$ iff $x \in x$. But we can prove a more general result. Let $x \in^n y$ say that there is an $(n-1)$-tuple $\vec{z}$ such that $x \in z_1 \in \ldots z_{n-1} \in y$. Then we have the following.

PROPOSITION 2.1.18 $\forall x(\mathcal{P}^{n}`x \subseteq x \rightarrow (\forall y \in^n y)(\mathcal{P}^{n}`(x - \iota`y) \subseteq (x - \iota`y)))$.

Proof: $\mathcal{P}^{n}`(x - \iota`y) \subseteq \mathcal{P}^{n}`x \subseteq x$. So $\mathcal{P}^{n}`(x - \iota`y) \subseteq (x - \iota`y)$ as long as $y \notin \mathcal{P}^{n}`(x - \iota`y)$. But $(y \in^n y) \rightarrow (\forall z)(y \in z \rightarrow y \in \bigcup^n z)$, so $(y \in^n y \wedge y \in \mathcal{P}^{n}`(x - \iota`y)) \rightarrow y \in x - \iota`x$. ∎

From this we have the following.

PROPOSITION 2.1.19 $(\forall x \in^n x)(\exists y)(\mathcal{P}^{n}`y \subseteq y \wedge x \notin y)$.

We will also need

PROPOSITION 2.1.20 $\forall x(\mathcal{P}^{n}`x \subseteq x \rightarrow (\exists y \subseteq x)(\mathcal{P}`y \subseteq y))$.

Proof: The desired witness is simply $x \cap \mathcal{P}`x \cap \ldots \cap \mathcal{P}^{n}`x$. This shows that *WF* is included in the intersection of all x such that $\mathcal{P}^{n}`x \subseteq x$ and enables us to conclude that, for example, if $x$ is well-founded then $x \notin^n x$ for all $n$. It follows from proposition 2.1.19 and proposition 2.1.20 that if $x \in^n x$ there is $y$ such that $x \notin y$ and $\mathcal{P}`y \subseteq y$, so $\neg WF(x)$. This is not otherwise obvious: all that is immediately obvious from the definition is that we can do induction for stratified formulae. ∎

Another surprising result is the following.

PROPOSITION 2.1.21 $WF \subseteq \mathtt{I} \cup \mathtt{II}$.

Proof: It is perfectly obvious that, if x is well-founded, then I or II must Win G_x. What is rather surprising is that there is a version of this highly unstratified allegation that can be proved in *NF*. By the definition of I (resp. II) as the intersection of all sets closed under $b \circ \mathcal{P}$ (resp. $\mathcal{P} \circ b$), it will be sufficient to show that, if $WF(x)$ and $\mathcal{P}`b`X \subseteq X$ and $b`\mathcal{P}`Y \subseteq Y$, then $x \in X$ or $x \in Y$. So by proposition 2.1.20 it will be sufficient that $\mathcal{P}^{n}`(X \cup Y) \subseteq (X \cup Y)$ for some $n$. Now in general $\mathcal{P}`(u \cup v) \subseteq \mathcal{P}`u \cup b`v$, so we have $\mathcal{P}`(X \cup Y) \subseteq \mathcal{P}`Y \cup b`X$ and $\mathcal{P}`(b`X \cup \mathcal{P}`Y) \subseteq (\mathcal{P}`b`X \cup b`\mathcal{P}`Y) \subseteq (X \cup Y)$ by hypothesis. So $\mathcal{P}^{2}`(X \cup Y) \subseteq (X \cup Y)$ as desired. ∎

PROPOSITION 2.1.22 *Every well-founded set is fixed by every (inner) automorphism of* $\langle V, \in \rangle$.

Proof: For any permutation σ of V at all, $\mathcal{P}`(\text{fix } \sigma) \subseteq (\text{fix } (j`\sigma))$ (fix σ is the set of points fixed by σ). Now, if σ is an automorphism of $\langle V, \in\rangle$ then $\sigma = j`\sigma$, so $\mathcal{P}`(\text{fix } \sigma) \subseteq (\text{fix } \sigma)$ and $WF \subseteq (\text{fix } \sigma)$ as claimed. ■

Again, although this result is standard in *ZF*, it merits comment that it is provable in *NF*, since it is unstratified. We cannot hope to remove the "(inner)" from the statement of this proposition, because of the logical possibility of (internally) well-founded sets that are moved by (external) automorphisms. There are after all many models of *ZF* plus the axiom of foundation that have external automorphisms!

REMARK 2.1.23 *There is no $\subseteq$-minimal set x such that $\mathcal{P}`x \subseteq x$.*

Proof: If $\mathcal{P}`x \subseteq x$ then $x \in x$, so by proposition 2.1.18 $\mathcal{P}`(x - \iota`x) \subseteq (x - \iota`x)$. ■

COROLLARY 2.1.24 *WF is a proper class.*

At present it is not known for which formulae one can do induction over *WF*. How much replacement *WF* satisfies is a mystery. No one has managed to prove that every well-founded set is included in a transitive well-founded set, let alone a minimal one, though there are no known strong consequences that might make it unprovable. It is tempting to suppose that one might be able to add a scheme of $\in$-induction for stratified formulae over *all* sets, but this temptation is easily banished by considering the (stratified) property of being distinct from V.

How large can well-founded sets be? We do not know very much about this either. On the one hand, there is no obvious way of proving that there cannot be large ($\overline{\overline{V}}$ sized) well-founded sets. If we could prove that $\mathcal{P}`\{y : \overline{\overline{y}} \leq T\ \overline{\overline{V}}\} \subseteq \{y : \overline{\overline{y}} \leq T\ \overline{\overline{V}}\}$ we would know that every set in *WF* is the size of some set of singletons. This looks plausible, but even to prove that there are at most $T\ \overline{\overline{V}}$ *pairs* seems to need AC_2. It is true that $\mathcal{P}`\{y : \overline{\overline{y}} \leq T\ \overline{\overline{V}}\} \subseteq \{y : \overline{\overline{y}} \leq T\ \overline{\overline{V}}\}$ is apparently stronger than we need, and that for any k

$$(\forall x)((\forall y \in^{\leq k} x)(\overline{\overline{y}} \leq \alpha) \rightarrow \overline{\overline{x}} \leq \alpha)$$

("$y \in^{\leq k} x$" is short for "$y \in x \ \vee\ y \in^2 x \ \vee \ldots \vee y \in^k x$") is sufficient to show that all *WF* sets are of size at most α, but I know of no choice of α or k for which this appears to be provable.

On the other hand we shall see in section 3.1.2 that it is even consistent relative to any stratified extension of *NF* for there to be a (non-standard) finite bound on the size of well-founded sets. If all well-founded sets are all finite, then either there are well-founded sets of all finite sizes or the upper bound is non-standard, for each member of V_ω can be constructed by hand. We shall see in section 3.1.1 that the existence of V_ω is consistent if AxCount$_\leq$ is. It is not known whether the existence of an infinite well-founded set is too strong to be relatively consistent with respect to *NF*.

No amount of ingenuity expended on proving unstratified facts about *WF* by induction will solve the question of what kind of well-founded set theory *WF* models. For example, *ZF* fails to decide $V = L$ even though it has all the unstratified induction one could want. However unlikely it may seem, it is not obviously out of the question that in *NF* there should be definable classes which, should they be sets, would have to be non-constructible. There are certainly definable subclasses of $\mathbb{N}$ that might not be sets—indeed there any many such examples: the class of all $n \in \mathbb{N}$ such that $n = Tn$ is one, but mostly these are not interesting; they are simply by-products of the failure of *NF* to reproduce all of unstratified analysis. However, in section 2.2, we shall see an example of a definition of a subclass of $\mathbb{N}$ that uses as parameters the kind of big cardinals that occur only in *NF*, and in contrast these definitions are potentially interesting.

Section 2.5 contains more on interpreting well-founded set theories into extensions of *NF*. In section 2.3.4 we will encounter a system (KF) that the well-founded sets in *NF* naturally model. We close with some rather speculative material on the closure properties of *WF*.

What do we know about the theory of well-founded sets in *NF*? Well, $WF \subseteq_e^{\mathcal{P}} V$, so $WF \models$ every $\Pi_1^{\mathcal{P}}$ theorem of *NF*, but this goes for any $X \subseteq_e^{\mathcal{P}} V$. *WF* is the directed intersection of all sets $X \supseteq \mathcal{P}`X$. Because we have the axiom of complementation, this tells us that $-WF$ is a directed union of all sets $X$ such that $X \subseteq b`X$. There is a theorem about sentences preserved under directed unions, so we can prove the following.

Remark 2.1.25 *Let Φ be $\Delta_0^{\mathcal{P}}$. If for arbitrarily small $X \supseteq \mathcal{P}`X$ we have*

$$(\forall \vec{x})[(\forall \vec{y})(\Phi(\vec{x},\vec{y})^X \to \bigvee_i y_i \in X) \to \bigvee_j x_j \in X]$$

then

$$\forall \vec{x}[(\forall \vec{y})(\Phi(\vec{x},\vec{y})^{WF} \to \bigvee_i y_i \in WF) \to \bigvee_j x_j \in WF].$$

Proof: Assume the premiss. Then by taking complements, we can see that, for arbitrarily large $X \subseteq b`x$, we have

$$(\forall \vec{x} \in X)(\exists \vec{y} \in X)(\neg\Phi(\vec{x},\vec{y})^{-X}).$$

We want to show

$$(\forall \vec{x} \in -WF)(\exists \vec{y} \in -WF)(\neg\Phi(\vec{x},\vec{y})^{WF}).$$

Let $\vec{x}$ be an n-tuple of things not in WF. Then, by directedness, there is some $X \subseteq b`x$ so that $\vec{x} \in X$. So, since $X \subseteq -WF$, $\exists \vec{y} \in -WF$ such that $\neg\Phi(\vec{x},\vec{y})^{-X}$. Now, since $\Phi \in \Delta_0^{\mathcal{P}}$, the truth value of $\neg\Phi(\vec{x},\vec{y})^{-X}$ is the same in any $-X$ such that $WF \subseteq_e^{\mathcal{P}} -X \subseteq_e^{\mathcal{P}} V$. But $X \subseteq b`X$, and so $-X \supseteq \mathcal{P}`-X$ and $WF \subseteq_e^{\mathcal{P}} -X \subseteq_e^{\mathcal{P}} V$ so $(\exists \vec{y} \in -WF)(\neg\Phi(\vec{x},\vec{y}))^{WF}$. That is,

$$(\forall \vec{x} \in -WF)(\exists \vec{y} \in -WF)(\neg\Phi(\vec{x},\vec{y}))^{WF}$$

which contraposes to

$$(\forall \vec{x})[(\forall \vec{y})(\Phi(\vec{x},\vec{y})^{WF} \rightarrow \bigvee_i y \in WF) \rightarrow \bigvee_j x_j \in WF].$$

■

In particular, if $\Phi(x,y)$ says that $y \in x$ we conclude unsurprisingly that every set of WF sets is WF.

There is a feeling that WF ought to be a substructure of V elementary for some sensible class Γ. What is Γ? Well, it is not Σ_1^{Levy} because of '$(\exists x)(x \in x)$'. While $str(\Sigma_1^{\mathcal{P}})$ is not obviously impossible, it certainly implies that there is an infinite well-founded set because the axiom of infinity is $str(\Sigma_1^{\mathcal{P}})$. However, the existence of an infinite set in WF seems to require $\text{AxCount}_{\leq}$ or something like it. WF certainly does not satisfy the $\forall_2$ fragment of *NF*, for consider '$\forall x \exists y(x \in y \in y)$'. '$WF \models str(\Pi_2^{\mathcal{P}} NF)$' seems to be the strongest conjecture that is not obviously false. Since $\neg AC$ is in $str(\Pi_2^{\mathcal{P}})$[19] and is provable in *NF*, this would imply that there is a well-founded set that cannot be well-ordered, and hence the axiom of infinity for well-founded sets. Since there can be a finite bound on the size of well-founded sets (see proposition 3.1.16) this is obviously not provable in *NF*. We shall see more of this discussion following theorem 3.1.28 in section 3.1.1.2.

2.2 Cardinal and ordinal arithmetic

As the reader will by now have seen, cardinals in *NF* have to be Russell–Whitehead cardinals. (S)he has probably guessed that ordinals have to be Russell–Whitehead ordinals. That is to say, the ordinal number $No`\langle X,R\rangle$ of a well-ordering $\langle X,R\rangle$ is the set of all well-orderings order-isomorphic to $\langle X,R\rangle$. *NC* is the set of all cardinals, *NO* the set of all ordinals. '*WC*' in this context has come to denote the set of all alephs.

The best way to approach the peculiarites of ordinal arithmetic in *NF* is to consider the Burali-Forti paradox concerning the order-type of the collection of all ordinals in their natural ordering. There is a more general version

[19] It is in fact $\Sigma_1^{\mathcal{P}}$.

which involves isomorphism classes of well-founded extensional relations. This version has never been discussed in print (as far as I am aware) and there is an excellent reason for this, namely that because of Mostowski's collapse lemma, it does not apparently serve to illustrate any point that is not made by the standard version with ordinals. Accordingly we will not discuss it. Interested readers may consult Hinnion's thesis [1975], after having considered remark 2.5.1.

We will start with a banality whose proof is probably less familiar than it should be.

LEMMA 2.2.1 *Every set of ordinals is naturally well-ordered.*

Proof: The obvious ordering places an ordinal α below β if any well-ordering of order-type α is isomorphic to a proper initial segment of any well-ordering of order-type β.

Let X be a set of ordinals, and $Y \subseteq X$. We must show that Y has a least member. Let us think of a member α of X and a well-ordering R of order-type α. To each non-zero member α' of Y with $\alpha' < \alpha$ we can associate an element of (the domain of) R: send α' to the R-sup of the (unique) initial segment of R of order-type α'. This is order-preserving, and is defined on an initial segment of Y. R is a well-ordering, so the range has a least element, and the preimage of this least element is the least element of Y. (The least element we have constructed actually has R as a parameter but, because of the existence of canonical isomorphisms between well-orderings, we can show that the result does not depend on the choice of R or α.)

The natural ordering is clearly total, since there is no difficulty in proving by induction that, given any two well-orderings, one is isomorphic to an initial segment of the other. ∎

It is very important that this proof does not depend on how ordinals are, so to speak, *implemented* set-theoretically.

Hartogs' theorem asserts that, for every set x, there is an ordinal too large to be the order-type of a well-ordering of a subset of x. The proof of Hartogs' theorem depends on the identity between (i) the least ordinal not representable as the order-type of a well-ordering of a subset of x, and (ii) the order-type of those ordinals in their natural ordering. This deserves close attention. To every proper initial segment X of the ordinals, we can associate *two* ordinals: the *least ordinal not in* X, and the *order-type* of X. Let us call them $L_1`X$ and $L_2`X$. These two ordinals belong to different types.[20] $L_1`X$ is clearly bigger than every member of X but this

[20]The real trouble with a youth spent gaining familiarity with von Neumann ordinals is that this very familiarity tends to destroy the type-theoretic intuitions necessary to appreciate other treatments of ordinals and this ignorance is then transmitted to the victim's students: in this respect, belief in von Neumann ordinals is rather like

is not obviously true of L_2. To prove the Burali-Forti paradox we would want to show $(\forall X)(L_1`X = L_2`X)$, and we would naturally want to do this by induction on the end-extension relation. The trouble is, since this allegation is unstratified, the class of counterexamples is not guaranteed to be a set, and so we cannot argue that it must have a minimal member. The conclusion is that there are initial segments X of the ordinals for which $L_1`X \neq L_2`X$. In fact it will usually turn out to be the case that if anything is to go wrong, then $L_1`X > L_2`X$. It is worth the reader's while to spend some time thinking about this, for some distinguished logicians have been trapped by inattention to this detail. Quine, in the (here uncited) first edition of [1951a] (published in 1940), adds a scheme of impredicative class existence to *NF* and restricts set existence to stratified expressions as before, with the intention of producing a theory that is related to *NF* in the way *GB* (with an impredicative class existence scheme) is related to *ZF*. However, he makes the mistake of allowing *bound class variables* into the set existence scheme; this means that the set abstracts of the new theory in the new language-with-classes are not just the old *NF*-terms, as would be the case if our set existence scheme required restriction of bound variables to sets, but new ones as well. In particular, consider the total orders R such that every sub*class* of the domain of R has a least element. This is clearly a stratified condition, and the collection of isomorphism classes of them becomes a *set* in the new theory. It comprises those ordinals over which we can perform any induction whose set of putative counterexamples corresponds to a class abstract. (Note that we are not now taking ordinals to be isomorphism classes where the *isomorphisms* may be proper classes, though that would be allowed too.) In particular, taking these objects to be *the* ordinals we can prove $(\forall X)(L_1`X = L_2`X)$ and thence the Burali-Forti paradox. This was noted by Lyndon and by Rosser [1942]. The corrected version of the theory (where set abstracts may not contain bound class variables) first appears in Wang [1950]. It is now known as "*ML*".

For further discussion of this point see section 2.2.1.

REMARK 2.2.2

$$\forall \alpha \in NO \;\; L_2`\alpha = T^2 L_1`\alpha.$$

This is stratified, and we can prove it by induction. ∎

One distasteful consequence of this is that, for very big ordinals, for example $\Omega = No`\langle NO, \leq\rangle$, we have $\Omega > T\Omega > T^2\Omega > \ldots$, so there is

child abuse—one generation's victims become the next generation's criminals. For a *ZF*-ist to understand Russell–Whitehead ordinals, (s)he must become a child again. "Brainwashed" is perhaps not too strong a word for what goes on. True, it has loaded resonances which serve my polemical purpose quite well, but it does have a precise core meaning which is appropriate here.

a sub*class* of NO with no least member. Since this subclass is countable, it means that, for some sense of "ω-standard" at least, no model of NF is ω-standard. Probably the most illuminating way to think of this is to acknowledge that the restricted nature of *separation* for NF means that certain big things which oughtn't to be classed as well-orderings come to appear to be well-ordered because we do not have enough separation to turn their subclasses-without-least-members into sets. See Rosser and Wang [1950].

THEOREM 2.2.3 *There are alephs too big to be the size of any set of singletons.*

Proof: For the moment let 'α' be a variable ranging over alephs, and suppose the theorem false. Then $(\forall\alpha)(\alpha \leq T\ \overline{\overline{V}})$. If there were $\alpha \not\leq T^2\ \overline{\overline{V}}$, then $T^{-1}\alpha$ would be a counterexample (a witness to the theorem) so $(\forall\alpha)(\alpha \leq T^2\ \overline{\overline{V}})$. Similarly $(\forall\alpha)(\alpha \leq T^3\ \overline{\overline{V}})$ and $(\forall\alpha)(\alpha \leq T^k\ \overline{\overline{V}})$ for any concrete k.

Now let X be the set of ordinals that are the order-types of well-orderings of subsets of ι^3"V, with their associated natural ordering. L_2'X certainly exists, and if we can show that T^{-2} of this ordinal exists, then it will be the sup of X and there will be a contradiction because, by the argument of the preceding paragraph, X must be NO. But the argument of the preceding paragraph also shows that T^{-2} of *any* ordinal must exist, since any ordinal ζ is the order-type of a well-ordering $\langle\iota^2$"$X, RUSC^2(R)\rangle$ of some subset of ι^2"V and then No'$\langle X, R\rangle$ is precisely T^{-2}'ζ. ■

What this tells us is that, for the biggest alephs at least, $T\aleph < \aleph$ and, in particular, that $\overline{\overline{NO}} \not\leq T^2\ \overline{\overline{V}}$.

Remark 2.2.2 tells us that any aleph $<\overline{\overline{NO}}$ is T^2 of something, and theorem 2.2.3 says that $\overline{\overline{NO}}$ itself is not T^2 of anything. The notation '$\aleph(\alpha)$' denotes the least aleph not $\leq \alpha$, Hartogs' aleph function. Thus $\overline{\overline{NO}} = \aleph(T^2\ \overline{\overline{V}})$ (so Ω is an initial ordinal). Now Hartogs' aleph function commutes with T so $T^{-1}\ \overline{\overline{NO}}$ should exist and be equal to $\aleph(T\ \overline{\overline{V}})$. Now $\aleph(T\ \overline{\overline{V}})$ certainly exists by theorem 2.2.3 so we have proved the following theorem.

THEOREM 2.2.4 $\overline{\overline{NO}} \leq T\ \overline{\overline{V}}$ *but* $\overline{\overline{NO}} \not\leq T^2\ \overline{\overline{V}}$.

■

(Of course the exponent (2 in this case) depends on the implementation of ordinals that we have chosen. It could equally well have been $\overline{\overline{NO}} \leq T^2\ \overline{\overline{V}}$ but $\overline{\overline{NO}} \not\leq T^3\ \overline{\overline{V}}$.) The failure of stratification that prevents us proving the

Burali-Forti paradox also prevents us proving a straightforward version of Hartogs' theorem. For any set X we can certainly construct the well-ordering (in their natural order) of the order-types of well-orders of subsets of X. L_1 of this well-ordering would certainly be longer than any of them, but we do not know that it is defined. L_2 of it certainly exists, but it may be smaller.

We can provide a definition of a rank function ρ on cardinals whose effect will be that $\rho`\alpha = 0$ if $\alpha$ is not $2^\beta$ for any $\beta$, and $= sup\{(\rho`\beta)+1 : 2^\beta = \alpha\}$ otherwise. $\rho`\alpha$ will be the rank of the tree $\mathcal{T}`\alpha$ (on page 30) in some sense of rank which I shall leave vague for the moment.

LEMMA 2.2.5 *Every well-founded relation has a rank.*

Proof: If R is a well-founded relation, we can construct things by recursion on it. In particular, we can construct a binary relation $S(x, y)$: "the rank of x is less than or equal to the rank of y" by recursion on R as follows: all minimal elements are related to each other by S; and recursively if every R-predecessor of x is S-related to some R-predecessor of y then $S(x, y)$. $S \cap S^{-1}$ is an equivalence relation. We prove by R-induction that any elements in its domain are S-comparable and that therefore the equivalence classes are totally ordered. They inherit well-foundedness from R. This well-ordering has an ordinal which is the rank of R. Although we will not need it here, a few words are in order on how we can also define, for any well-founded structure, a rank function on its elements.

Let $*R$ be the union of all finite iterates of R. Then R restricted to $*R``(\iota`x)$ is another well-founded relation and has a rank as above. This can can be thought of as pertaining to the element x in the domain of R, so that we have a rank function defined on everything in the domain of R. ■

Now let us define $\rho`\alpha$ to be the rank of $\mathcal{T}`\alpha$. This will not be homogeneous, but it is stratified. For this to be legitimate, we have to prove that $\mathcal{T}`\alpha$ is a well-founded structure.

THEOREM 2.2.6 *$\mathcal{T}`\alpha$ is well-founded.*

Proof: Suppose it is not; then there is a set X of cardinals $\subseteq \mathcal{T}`\alpha$ such that $(\forall\beta \in X)(\exists\gamma \in X)(2^\gamma = \beta)$. Let us call members of $X$ *unranked.* $X$ is a subtree of $\mathcal{T}`\alpha$. Let $[\alpha]_0 = \iota`\alpha$ and $[\alpha]_{n+1} = \{\beta : 2^\beta \in [\alpha]_n \wedge \beta \text{ unranked}\}$ for other n. Let β_n be inf $\aleph``[\alpha]_n$, where $\aleph$ is Hartogs' aleph function. By a standard result of Sierpinski, $\aleph`\alpha < \aleph`2^{2^{2^\alpha}}$ and so $\beta_{n+3} < \beta_n$, which is impossible since the β_n are alephs. ■

The fact that there can be no infinite descending chain in $\mathcal{T}`\alpha$ has the immediate consequence that the theory of negative types—although it is obviously consistent by a compactness argument and therefore has models—nevertheless has no true second-order models, that is to say, models where the $(n+1)$th type is genuinely the power set of the nth type (see Forster [1989]).

We will use ρ to prove the axiom of infinity.

THEOREM 2.2.7 Specker [1953]. *V is infinite.*

Earlier I said that 2^{α} is undefined if $\alpha \not\leq T\,\overline{\overline{V}}$. We can relax a little and allow 2^{α} to be defined as $T^{-1}2^{T\alpha}$ if $2^{T\alpha} \leq T\,\overline{\overline{V}}$ (see Crabbé [1982b] and [1984]). If we do this, then T is indeed an isomorphism between $\mathcal{T}\text{‘}\,\overline{\overline{V}}$ and $\mathcal{T}\text{‘}T\,\overline{\overline{V}}$. If we stick to the original version, then T is an isomorphism between $\mathcal{T}\text{‘}T\,\overline{\overline{V}}$ and $\mathcal{T}\text{‘}T^2\,\overline{\overline{V}}$. (For any cardinal α, we can prove by induction on $\mathcal{T}\text{‘}\alpha$ that $(\forall\beta \in \mathcal{T}\text{‘}\alpha)(\rho\text{‘}T\beta = T\rho\text{‘}\beta)$. This induction might not look stratified but it is, for ‘ $u = \rho\text{‘}v$ ’ is a stratified expression with only *two* free variables, not three.) While these details are irritating and have to be mastered, they affect only our choice of example, and do not perturb the underlying mathematics. Let us suppose we have defined exponentiation so that T is an isomorphism between $\mathcal{T}\text{‘}\,\overline{\overline{V}}$ and $\mathcal{T}\text{‘}T\,\overline{\overline{V}}$. If $\rho\text{‘}\,\overline{\overline{V}}$ is an infinite ordinal then we are done, so suppose $\rho\text{‘}\,\overline{\overline{V}}$ is finite. Since $2^{\overline{\overline{\iota\text{“}V}}} = \overline{\overline{V}}$, we know that $\rho\text{‘}\,\overline{\overline{V}} \geq 1+\rho\text{‘}\,\overline{\overline{\iota\text{“}V}}$. Now $Tn+1$ cannot be equal to n, since Tn is even iff n is (being even is stratified); so $n \geq Tn+1 \rightarrow n > Tn+1$ and in particular, $\rho\text{‘}\,\overline{\overline{V}} > 1+\rho\text{‘}\,\overline{\overline{\iota\text{“}V}}$. Therefore $(\exists\beta)(\beta \neq \overline{\overline{\iota\text{“}V}} \wedge 2^{\beta} = \overline{\overline{V}})$. Nothing in $\mathbb{N}$ can have two distinct base-2 logarithms, so V must be infinite. ■

It is worth spending some thought on this proof, for a refined version proves the powerful lemma of Henson below (lemma 2.2.10).

Specker's original proof—which he never published—is simple and memorable. We include it here because it has the distinctive flavour of the T operation and parity checks typical of proofs involving big cardinals, $\overline{\overline{V}}$, $\overline{\overline{NO}}$, and so on.

Suppose $\overline{\overline{V}}$ is finite. Then, for some finite n (to wit: $\overline{\overline{V}}$), $n = 2^{Tn}$. So n is an integer of the form $2^{2^{2^{\cdot\cdot}}}$, and so also is Tn. This is because the assertion that the tower-length function commutes with T is stratified and can be proved by induction.[21] If the height of the tower associated with n is k, then the height of the tower associated with Tn is Tk. But since

[21] Though the assertion that f commutes with T, with ‘f’ free, is *un*stratified. There is no problem here because the f we are interested in is a closed term, and therefore the thing we are trying to prove can be rewritten to have no free variables other than the integers.

$n = 2^{Tn}$, we must have $k = Tk + 1$ which is impossible: k is divisible by 2 iff Tk is divisible by $T2$, which is 2. ■

THEOREM 2.2.8 Specker [1953]. $NF \vdash \neg AC$.

Proof: We will assume AC and derive a contradiction. We noted at the beginning of section 2.1 that 2^α is defined iff $\alpha \leq T\ \overline{\overline{V}}$. Let us stick to this definition for the moment. We noted also that the relation "$\beta = 2^\alpha$" is *homogeneous.* This has the consequence that, for any cardinal α, the collection $\{\alpha, 2^\alpha, 2^{2^\alpha} \ldots\}$ is a *set*, for we can define it by the stratified set abstract $\bigcap\{X : \alpha \in X \wedge (\forall \beta \in X)(2^\beta \text{ exists} \rightarrow 2^\beta \in X)\}$. Specker writes this "$\Phi`\alpha$" in [1953] and this is now standard notation. It also means that we can form the set (which in Forster [1976, 1983a] I call "*SM*") of those cardinals $\alpha$ for which $\Phi`\alpha$ is finite. We prove by induction on $n \in \mathbb{N}$ that T of the nth member of $\Phi`\beta$ is the $Tn$th member of $\Phi`T\beta$. This is permissible because there is a stratified expression $F(n, \alpha, \beta)$ which says that β is the nth member of $\Phi`\alpha$. Therefore, if $\Phi`T\kappa$ is finite, so is $\Phi`\kappa$. By $AC$, $SM$ must have a minimal member, $\alpha$ say. We will be interested in the size $n$ of $\Phi`\alpha$. Let us consider the maximal member β of $\Phi`\alpha$. Since $2^\beta$ is not defined, $\beta \not\leq T\ \overline{\overline{V}}$. Therefore, by $AC$, $T\ \overline{\overline{V}} < \beta$. Evidently $T\beta$ is the $Tn$th member of $\Phi`T\alpha$, so $\Phi`T\alpha$ is $T``\Phi`\alpha \ \cup\ \Phi`T\beta$. Now what is $\Phi`T\beta$? Well, we know $T\ \overline{\overline{V}} < \beta \leq \overline{\overline{V}}$ so $T^2\ \overline{\overline{V}} < T\beta \leq T\ \overline{\overline{V}}$ and $T\ \overline{\overline{V}} = 2^{T^2 \overline{\overline{V}}} \leq 2^{T\beta} \leq 2^{T \overline{\overline{V}}} = \overline{\overline{V}}$.

Therefore $\Phi`T\beta$ contains either one (if $2^{T\beta} > T\ \overline{\overline{V}}$) or two (if $2^{T\beta} = T\ \overline{\overline{V}}$) new members ($T\beta$ is already in $T``\Phi`\alpha$), so the size of $\Phi`T\alpha$ is either $Tn + 1$ or $Tn + 2$. What we do next is to show that $\alpha = T\alpha$, so that $n = Tn+1$ or $Tn+2$, either of which is impossible.[22] Now we have already shown that $\Phi`T\alpha$ is also finite,[23] so $\alpha \leq T\alpha$ by minimality of $\alpha$, and we know anyway that $\Phi`\kappa$ is finite if $\Phi`T\kappa$ is, so $\alpha \leq T^{-1}\alpha$ and therefore $\alpha = T\alpha$ as desired. ■

It is important to appreciate that this refutation of AC depends on the use of very big cardinals, and so it does not generalize to refutations of versions of AC which are restricted to low sets. It gives us no reason to suppose that AC will fail in WF, for example. Somewhat surprisingly, full AC seems to be the *only* natural choice principle which we know how to refute in *NF*. On the other hand remark 2.2.9 shows, however, that quite a weak (though unnatural) version of AC is refutable.

REMARK 2.2.9 $\overline{\overline{V}}$ *is not a beth number.*

[22] It is easy to show that if $n > Tn$ then every standard integer divides $n - Tn$.

[23] Beware the misprint in Henson's otherwise limpid exposition [1973a] where '$\Phi(n)$' and '$\Phi(Tn)$' have been permuted in the statement of lemma 2.1 (a), top of p. 62.

Proof: As we have seen, the assertion '$\alpha = \beth_\zeta$' is stratified and will commute with T. Thus, if $\overline{\overline{V}}$ were $\beth_\zeta$, then $T\ \overline{\overline{V}}$ would be $\beth_{T\zeta}$; but $\overline{\overline{V}} = 2^{T\overline{\overline{V}}} = 2^{\beth_{T\zeta}} = \beth_{T\zeta+1}$, so $\zeta = T\zeta + 1$. This is impossible as before. ∎

However, we do know (since $\overline{\overline{V}} = 2^{T\overline{\overline{V}}}$) that, for each standard n, there is α so that $\overline{\overline{V}} = \beth_n\text{`}\alpha$. Thus we can "descend" from $\overline{\overline{V}}$ by beth numbers, and "ascend" from $\aleph_0$, though we can never meet in the middle!

For Henson's lemma below, we will need a bit more notation. Let us write "$incr(f)$" to mean that f is a non-decreasing function on a proper initial segment of NO, where the least thing not in the domain is to be notated "$\Theta\text{`}f$". By 'f^T' we mean '$\{\langle T\alpha, T\beta\rangle : \langle\alpha,\beta\rangle \in f\}$'.

LEMMA 2.2.10 Henson [1973a]. *If $incr(f)$ and $T\text{`}\Theta\text{`}f < \Theta\text{`}f$ and $\Theta\text{`}f \leq f\text{`}T(\Theta\text{`}f)$ and f commutes with T, then there is an ordinal α such that $\alpha = T\alpha$ and $\Theta\text{`}f < f\text{`}\alpha \leq f\text{`}T(\Theta\text{`}f)$.*

Proof: We will need another definition to supply the proof. Let $\Phi(f,\alpha)$ be the closure of $\iota\text{`}\alpha$ under f. The notational evocation of Specker's Φ function is deliberate. By an induction of the kind that it should now be safe to leave as an exercise, we note that $incr(f)$ implies that $\Phi(f^T, T\alpha) = T\text{``}\Phi(f,\alpha)$. The proof of the main result now follows quite closely Specker's refutation of AC.

Fix f. Then let Θ_0 be the least α such that $\Phi(f,\alpha)$ is finite. The first two conditions in the antecedent of this lemma ensure that f^T is the restriction of f to $\{\alpha : \alpha < T\text{`}(\Theta\text{`}f)\}$, and it follows easily that $\Theta_0 = T\Theta_0$. Now let β_0 be the last member of $\Phi(f,\Theta_0)$. Then, as in Specker's proof, $\Phi(f, T\beta_0)$ has one or two members and, since $n \neq Tn + 1$ always, we must have $\Phi(f,\Theta_0) = T\text{``}\Phi(f,\Theta_0)$. Therefore $\beta_0 = T\beta_0$, and the α whose existence is claimed in the statement of the theorem is the least ζ such that $f\text{`}\zeta = \beta_0$, which is certainly going to be cantorian. ∎

Henson states, but does not prove, the following analogue of this.

LEMMA 2.2.11 Henson [1973a]. *If $incr(f)$ and $T\text{`}\Theta\text{`}f > \Theta\text{`}f$ and $T\Theta\text{`}f \leq f^T\text{`}(\Theta\text{`}f)$ and f commutes with T, then there is an ordinal α such that $\alpha = T\alpha$ and $T\Theta\text{`}f < f\text{`}\alpha \leq f^T\text{`}(\Theta\text{`}f)$.*

The existence of $\aleph_\omega$ is a problem for *NF* just as it was for Russell and Whitehead, and for the same reasons. For each concrete integer 1, 2, 3, ..., we can prove *by hand* the existence of the aleph with the corresponding subscript. Let us see what happens if we try the obvious recursive construction of a set of size $\aleph_{n+1}$ from a set of size $\aleph_n$. Suppose X is of size $\aleph_n$. Then clearly the set of all isomorphism classes of well-orderings of X should be of size $\aleph_{n+1}$. However, it is two types higher, and so what this construction gives us is not a proof of "there is a thing of size $\aleph_n$" → "there is a thing of size $\aleph_{n+1}$", but rather "there is a thing of size $\aleph_n$" $\rightarrow$ "there is a thing of size $\aleph_{T^2 n+1}$". We have exactly the same problem about $\beth_n$ with $n \in \mathbb{N}$.

If we assume AxCount$_{\leq}$, then we can prove the existence of $\aleph_{n+1}$ and $\beth_{n+1}$ for all $n \in \mathbb{N}$ quite simply, because (as we have seen in section 2.1.1) AxCount$_{\leq}$ is equivalent to the assertion that $Tn < m$ is well-founded on $\mathbb{N}$.

In the absence of AxCount$_{\leq}$, as Specker has pointed out, we apparently do not even know that the set of infinite cardinals, NCI, is infinite. Despite a lot of effort, no contradiction has been derived from the assumption that NCI is finite. The best we can do is show that since it implies that any monotone 1-1 $f : NCI \rightarrow NCI$ is the identity, then $\forall \alpha \in NCI \ \alpha = 2\alpha$, since multiplication by 2 is known to be monotone and 1-1. This easily shows that, in these circumstances, $\langle NCI, \leq \rangle$ is a distributive lattice. When we reflect that in Z we clearly cannot prove the existence of an infinite set of infinite cardinals, then the intractability of this problem for *NF* is an indication that plain *NF* (without AxCount) is perhaps no stronger than Z. Even so, *NF* might conceivably prove that there is an infinite set of infinite cardinals. The grounds for this hope are as follows: in *NF* (since *NCI* is a set) the negation (*NCI* finite) puts a *finite* bound on the size of sets of infinite cardinals.[24] If there are few cardinals, many sets have to be the same size and we can infer choice principles—as we have seen, if *NCI* is finite then $\forall \alpha \in NCI \ \alpha = 2\alpha$—and choice principles are very threatening to the consistency of *NF*.

A natural question is: what is the cardinality[25] of *NO*? Since we have not proved that NCI is infinite, nor *a fortiori* have we proved that $\overline{\overline{NO}}$ is limit. If we have AC_{wo} (well-ordered sets have choice functions), we can actually *prove* that $\overline{\overline{NO}}$ is a successor cardinal.

Remark 2.2.12 *AC_{wo} implies that $\overline{\overline{NO}}$ is a successor aleph.*

Proof: Assume $\overline{\overline{NO}}$ is limit. *NO* is well-ordered, so pick one element $\langle X_\alpha, R_\alpha \rangle$ from each α. Take the union of all the X_α, and well-order it by $x < y$ if x first appears earlier than y or if they appear first in the same X_α and $x R_\alpha y$. This is clearly a well-ordering of $\bigcup_{\alpha \in NO} X_\alpha$ and must be at least as

[24] In contrast, the corresponding assertion in Z—that there are no infinite sets of infinite cardinals—is quite compatible with there being arbitrarily large finite sets of infinite cardinals. Similarly in *NF* the existence of a universal set blurs the distinction between finitism (everything is finite) and ultrafinitism (there are only finitely many things).

[25] 'Size' and 'cardinality' are synonymous.

long as any of the R_α, since $\overline{\overline{NO}}$ is limit. But then we have a well-ordering dominating all other well-orderings, which is impossible. ■

This is unexpected: one might have thought that there would be a large cardinal lurking somewhere behind $\overline{\overline{NO}}$, but it has not revealed itself. As we have seen, it appears that if the axiom of counting is false, and there are non-standard integers, then $\overline{\overline{NO}}$ might be $\aleph_n$ with n non-standard. Of course this cannot happen if the axiom of counting (or even AxCount) holds, so it seems that (even without resorting to consistency assertions in arithmetic) we can find unstratified combinatorial assertions about quite small sets that can compel $\overline{\overline{NO}}$ to be quite large. We shall see next how AxCount$_\leq$ implies a piece of *stratified* combinatorics whose consistency relative to *ZF* is not known.

PROPOSITION 2.2.13 *ρ‘$\alpha \geq \omega$ implies that α is not an aleph.*

Proof: Suppose not, and that α is an aleph of infinite rank. Let α_0 be the least element of $\mathcal{T}$‘α. Then, for some n, we have $\alpha_0 \in [\alpha]_n$. Now, since ρ‘$\alpha \geq \omega$, $[\alpha]_k$ is non-empty for all k, so consider an arbitrary $\beta \in [\alpha]_k$ with $k > n$. Then $\alpha_0 \leq \beta$, $2^{\alpha_0} \leq 2^\beta$, and $\beth_n$‘$\alpha_0 \leq \beth_n$‘β; but $\beth_n$‘$\alpha_0 = \alpha$, and $\beth_n$‘$\beta < \alpha$ since $n < k$. ■

This proof will actually show something slightly stronger. Let $f(n, \alpha)$ be ρ‘($\beth_n$‘α) (ignoring the possibility that these beth numbers might not be defined). Obviously $(\forall n \forall \alpha) f(n, \alpha) \geq n$. Now, if *AC* holds, $f(n, \alpha)$ is eventually $n + k$ for some constant k. Otherwise we obtain a contradiction in the above style by considering the least β such that, for some m, n, $\beth_n$‘$\beta = \beth_m$‘α.

We can use proposition 2.2.13 to prove an observation of Forster [1985]. Theorem 3 of that paper states that $TNT + AC$ has no ω-standard model. But suppose it did. Consider $\overline{\overline{V_0}}$ (that is to say, the cardinal of the universe at type 0). By *AC* it is of finite rank but, since $\overline{\overline{V_0}} = \beth_n$‘ $\overline{\overline{V_{-n}}}$ for each standard integer n, the rank of $\overline{\overline{V_0}}$ has to be non-standard.

Since ‘*NCI* infinite’ is still open, we evidently do not know how to prove in *NF* the existence of a cardinal of infinite rank. It is an open question whether the existence of such cardinals is consistent with *ZF*. There are plenty of cardinals in *NF* that *ought* to be of infinite rank: $\overline{\overline{V}}$ is one. Suppose we have defined exponentiation so that T is indeed an isomorphism between $\mathcal{T}$‘ $\overline{\overline{V}}$ and $\mathcal{T}$‘T $\overline{\overline{V}}$ (otherwise we use our original definition, and then T is an isomorphism between $\mathcal{T}$‘T^2 $\overline{\overline{V}}$ and $\mathcal{T}$‘T $\overline{\overline{V}}$). Recall our proof earlier of the axiom of infinity. We showed $T\rho$‘ $\overline{\overline{V}} = \rho$‘$T$ $\overline{\overline{V}}$. But obviously ρ‘ $\overline{\overline{V}} > \rho$‘ $\overline{\overline{\iota``V}}$ since $2^{\overline{\overline{\iota``V}}} = \overline{\overline{V}}$. So ρ‘ $\overline{\overline{V}} > T\rho$‘ $\overline{\overline{V}}$. If we have AxCount$_\leq$ then we infer that $\rho`\overline{\overline{V}}$ must be infinite. Since the existence of a cardinal of infinite rank is not known to be relatively consistent even with *ZF* this is in contrast to plain *NF*, which does not appear to be significantly stronger than $Z$. If $\alpha$ is infinite and $\rho`\alpha \geq 4$ then $\alpha = \alpha^2$, so a simple application of Sierpinski's remark that $\aleph`\alpha < \aleph`2^{2^{\alpha^2}}$ shows the following.

REMARK 2.2.14 *For some small fixed finite k,* $\aleph`T^k\alpha \geq \aleph_{\rho`\alpha}$.

Proof: Use induction on rank (at least if $\rho`\alpha$ is limit). Hence $\aleph_\zeta$ exists for any ordinal ζ which is below the rank of some cardinal. ■

In fact we can replace the sequence of alephs in this remark by the (possibly faster-growing) sequence ζ_α, defined for the nonce by $\zeta_0 = \aleph_0$ and $\zeta_{\alpha+1} = \aleph`2^{\zeta_\alpha}$ taking sups at limit α.

We saw at the outset that, in *NF*, $(\forall\alpha)(\alpha \leq T\alpha)$ is impossible. We shall see next that $(\forall\alpha)(T\alpha \leq \alpha)$ has strong consequences.

THEOREM 2.2.15 $(\forall\alpha \in NC)(T\alpha \leq \alpha) \rightarrow$ *AxCount.*

Proof: $(\forall\alpha)(T\alpha \leq \alpha)$ obviously implies AxCount$_{\geq}$ so it will be sufficient to derive AxCount$_{\leq}$. Consider $\mathcal{T}`\overline{\overline{V}}$. In general, if $\beta$ is in the $k$th level of this tree, $T\beta$ is in the $Tk$th level of $\mathcal{T}`T\overline{\overline{V}}$ and so in the $Tk+1$th level of $\mathcal{T}`\overline{\overline{V}}$. Since $(\forall\alpha \in NC)(T\alpha \leq \alpha)$, we have $\beta \geq T\beta$, so $\overline{\overline{\Phi`\beta}} \leq \overline{\overline{\Phi`T\beta}}$ and $k \leq Tk+1$. We know that $k \neq Tk+1$; so $k \leq Tk$ for any $k$ such that $\mathcal{T}`\overline{\overline{V}}$ has members at level k. Therefore, if we can show that $\mathcal{T}`\overline{\overline{V}}$ is of infinite rank, we will have proved AxCount$_{\leq}$. Suppose it is not; then there is a maximal $n$ such that $\mathcal{T}`\overline{\overline{V}}$ contains a cardinal β at that level. But then $n \leq Tn$ and then $T\beta$ is in $\mathcal{T}`\overline{\overline{V}}$ at level $Tn+1$ so n was not maximal. ■

We shall see later (theorem 3.1.35) that even $(\forall\alpha \in NC)(\alpha \leq^* T\alpha \lor T\alpha \leq^* \alpha)$ is not provable.

Specker's Φ function provides us with a way of defining sub*classes* of $\mathbb{N}$ which have no obvious set definition. We know that $\overline{\overline{\Phi`T\alpha}} \geq T(\overline{\overline{\Phi`\alpha}})+1$. We can associate with a big cardinal α and a (standard) integer n, the cardinal number of $\Phi`T^n\alpha$. This quantity may or may not be finite but, once it becomes infinite, it remains infinite for all larger n. In cases where it is finite for all n, we have associated to α an ω-sequence of integers whose definition is unstratified. In an obvious way, it corresponds to a (possibly proper) class of hereditarily finite sets. Since it codes information about big sets, there is no reason to suppose that it can also be generated by manipulations inside *WF*. This is much more interesting than the possibility mentioned earlier of the existence of finite proper classes of integers, since this construction does not depend on the failure of unstratified induction over $\mathbb{N}$. Indeed we cannot construct these objects as *sets* without some unstratified comprehension which *NF* apparently lacks: $\{T^{n}`\alpha : n \in \mathbb{N}\}$ cannot be a set for arbitrary α, for if $\alpha = NO$, we would obtain a set of alephs with no least member. There are other constructions of sets of ordinals, which, although they can be executed in *ZF*, will not necessarily have the same meaning. For any cardinal α, $\rho``(\Phi`\alpha)$ is a set. If $\Phi`\alpha$ is infinite, this will be an infinite set of (possibly infinite) ordinals. In ZFC this construction will always produce an increasing ω-sequence of integers but, in *NF*, because of the existence of big cardinals, these increasing ω-sequences might contain infinite ordinals and be highly non-constructible (in the sense of corresponding to non-constructible well-founded sets).

2.2.1 *Some remarks on inductive definitions*

The existence of big sets greatly simplifies the construction of inductively defined sets. If we want the closure of X under some n-ary (homogeneous) function f, this is simply $\bigcap\{Y : X \subseteq Y \wedge f``Y^n \subseteq Y\}$. This always works if $f$ is a set, simply because the set abstract always contains $V$ (which is closed under everything!). In *ZF*, by contrast, $\{Y : X \subseteq Y \wedge f``Y^n \subseteq Y\}$ is not a set, and we have to do the construction "locally", inside one such Y. The problem then is to show that there are such Y at all, and this is where we typically construct one by recursion over the ordinals. This sounds more straightforward than it is. If f is a function with infinitely many places, then there is no reason to suppose the process will converge in ω steps,[26] and we have to invoke (and first prove!) Hartogs' theorem so we can be confident that we will not run out of ordinals. By being able to construct inductively defined sets directly, *NF* is curiously more idiomatic (i.e. closer to the spirit of naïve set theory) than *ZF* is. In *ZF* the need for a set in which to perform "locally" the construction at the beginning of this paragraph is met by an axiom of infinity. This axiom is traditionally expressed by an assertion that there is a set that provides a natural implementation of arithmetic, namely the von Neumann ω. Thus in *ZF* the need to construct things by recursion over the integers or ordinals sets up as the centrepiece of set theory something which is, after all, merely an implementation of just another branch of mathematics (arithmetic).

There is a virtue in this compulsion *NF* is under to represent all constructions of inductively defined sets as such constructions *ab initio*, rather than piggy-backing on the ordinals or the naturals, for it concentrates our minds on the fact that no inductively defined set is logically prior to any other. There is also a necessity. In *NF* we do not know that there are enough ordinals! As we have seen (theorem 2.2.3), Hartogs' theorem fails for some very big sets: L_2 of the set of order-types of well-orderings of bits of x exists

[26]Even that is satisfactory only if we already know what $\mathbb{N}$ is!

for all x, but not necessarily L_1, which is what we would actually need. In *NF*, the chief practical implication of the fact that $\exists X(L_1`X > L_2`X)$ is that we do not define long well-orderings by recursion on the ordinals as we do in *ZF*, but must construct them as the intersection of all sets containing this and closed under that, for otherwise our construction is bounded by Ω, the order-type of $\langle NO, < \rangle$, which may be less than the thing we are after. I have argued above that this is more idiomatic. Be that as it may, it is certainly inescapable here!

To illustrate, I shall sketch my original proof of proposition 2.1.16. If the *TC* function is a set, then the function (call it "*D*" for the nonce) sending x to $\bigcup\{y : TC`y \subseteq x \ \wedge TC`y \neq x\}$ is likewise a set. We now construct a sequence $\langle V, D`V, D^2`V \ldots\rangle$ taking intersections at limits, using recursion on ordinals, which is legitimate because functional composition is homogeneous. It can be shown that this sequence never terminates and therefore is at least as long as any well-ordering, which is impossible, since there is clearly no last ordinal.[27] If we perform this construction by indexing over the ordinals, we stop when we run out of ordinals, and do not discover the contradiction since there are plenty of well-orderings longer than the ordinals in their natural order. To discover the contradiction we have to construct the whole sequence as "the intersection of all sets containing this and closed under that" and then ask questions about the ordinal number associated with this sequence.

Another example is the subject of a question of Boffa's. *WO* is the set of all sets that can be well-ordered. Define x' to be $\{\bigcup y : y \in WO \ \wedge \ y \subseteq x\}$ and then $\mathbf{W}$ to be the intersection of all sets extending *WO* and closed under $'$ and well-ordered unions. We have no prima facie reason to suppose that this sequence is shorter than the ordinals, so we cannot construct this by saying: let W_0 be *WO*, and let $W_{\alpha+1} = \{\bigcup x : x \in WO \wedge x \subseteq W_\alpha\}$ taking unions at limit ordinals. It is not hard to show that $\langle \mathbf{W}, \subseteq\rangle$ must be a well-ordering, and must therefore have an order-type. This order-type must be successor, for $\bigcup \mathbf{W}$ is a well-ordered union of things in $\mathbf{W}$ and therefore belongs to $\mathbf{W}$. So $x = x'$ for some x in $\mathbf{W}$.[28]

Another illustration is to be found in section 1.1.3. There H_X, the collection of things that are hereditarily in X, is defined in *NF* as $\bigcap\bigcap\{Y : X \in Y \wedge \mathcal{P}``Y \subseteq Y\}$, where the $\bigcap$ are taken over definable classes of sets. $\bigcap\{Y : X \in Y \wedge \mathcal{P}``Y \subseteq Y\}$ (one $\bigcap$ only!) is clearly intended to be the (proper) class $\{Y, \mathcal{P}`Y, \mathcal{P}^2`Y, \ldots\}$. If we try to construct *this* by

[27]Given a well-ordering of order-type α, peel an element off the beginning and put it on the end to get one of order-type $\alpha + 1$. Do not confuse this with the question of whether or not there is a last *initial* ordinal!

[28]Although we have used this construction here merely to illustrate the importance of constructing well-orderings properly, it is of some interest in its own right. Boffa's question is: is this fixed point V?

recursion over $\mathbb{N}$, we realize that the recursion step is unstratified, since $\mathcal{P}$ is inhomogeneous, and so it is not clear what we mean. The only way to make it clear to ourselves what we do mean is to construct the desired object directly by an *ab initio* inductive definition, which is what was done above.

2.2.2 *Closure properties of small sets*

Any definable class which is a non-principal ideal in $\langle V, \subseteq \rangle$ is a candidate for a notion of *smallness.* Some of these are familiar from other contexts—well-founded sets, well-ordered sets, totally ordered sets—and do not attract particular attention here. For us, the most important notions are those of cantorian and strongly cantorian sets. We saw some closure properties of these at the beginning of this chapter. These closure properties are not well understood, and a few comments are probably in order.

REMARK 2.2.16 *If x is a finite, strongly cantorian set of disjoint cantorian sets, then $\bigcup x$ is cantorian.*

Proof: If $\vec{u}$ and $\vec{v}$ are finite strings of the same length such that, for all entries, $\overline{\overline{u_i}} = \overline{\overline{v_i}}$, then clearly, by a use of AC for finite sets, we can pick a finite set of bijections $u_i \longleftrightarrow v_i$. If $\vec{y}$ is a finite strongly cantorian list of pairwise disjoint cantorian sets, then we are in the above situation. To see this, take $\vec{u}$ to be $\vec{y}$ and the v_i to be $\iota``y_i$. Then $v_{Ti} = \iota``u_i$ for all i. This happens whatever the length of the strings, simply because the $\vec{y}$ are cantorian. In this case, since the list is strongly cantorian, $i = Ti$ always, and so $v_i = \iota``u_i$. Now we use finite AC as before, and the union of the selected bijections is another bijection. ∎

For this particular proof to work, it is necessary that the index set be strongly cantorian. AC for strongly cantorian sets would suffice to show that a union of a strongly cantorian set of disjoint cantorian sets is cantorian. Obviously arbitrary unions of disjoint cantorian sets cannot be cantorian in general, since V is a union of singletons, but there seems to be no similarly obvious reason why a union of a cantorian family of disjoint cantorian sets should not be cantorian. It is worth pointing out that cantorian forms of the axiom of choice would not help here, because we need the set of indices to be strongly cantorian to ensure that, in picking a bijection between (as it were) u_i and v_i, we are in fact picking a map from u_i to $\iota``u_i$. Thus we need a comprehension axiom rather than a choice axiom. Similarly one might expect that a union of a strongly cantorian set of strongly cantorian sets is likewise strongly cantorian, but here again the problem is not one of lack of AC, for we know already what set it is we want; it is rather that there seems to be no stratified comprehension axiom we can use to prove its existence.

In section 2.2 we considered the possibility that NCI (the set of cardinals of infinite numbers) might be infinite. If this is false, then $(\forall\alpha \in NCI)(\alpha = 2\cdot\alpha)$ so (for infinite sets at least) unions are the same size as disjoint unions. In particular, a union of two infinite cantorian sets is cantorian. This looks sensible enough, but it might imply AxCount, or AxCount$_{\leq}$, in which case (since it is stratified) it would be refutable (corollary 2.3.10). Consideration of uniqueness of subtraction for natural numbers shows that, if the union of two finite cantorian sets is cantorian (which does not appear to follow from $(\forall\alpha \in NCI)(\alpha = 2\cdot\alpha)$), then every finite cantorian set would be strongly cantorian. Does this imply AxCount? If we knew that the only constraint on the restriction of T to $\mathbb{N}$ was that it should be an automorphism, then we would expect the answer to be "no", for there are non-standard models of arithmetic with an automorphism that fixes more than just the standard integers. We would need to know that the arithmetic of *NF* is a conservative extension of something sensible (we know this for *NF*$_3$: see Pabion [1980]) but this must await a consistency result for *NF* itself. It would also depend on the subclasses of $\mathbb{N}$ that we can define using big cardinals (as on page 55) not turning out to be sets not definable in (as it might be) analysis.

For each notion of smallness it is natural to ask if we can prove that there is an infinite set with that property. *NF* proves that there is an infinite well-ordered cantorian set. We shall see that if *NF* is consistent then it does not prove that there is an infinite strongly cantorian set (corollary 2.3.10) nor that there is an infinite well-founded set (see proposition 3.1.16). We know that the assertion that there is an infinite strongly cantorian set implies the consistency of *NF* but it seems to be open whether or not *NF* plus the existence of infinite well-founded sets is equiconsistent with *NF*.

There are no obvious connections between any of these notions of smallness: no reason to suppose that all well-founded sets are well-orderable or totally orderable or vice versa, no reason to suppose that cantorian sets are well-orderable, and so on. In [1973a] Henson considers an axiom asserting that all well-ordered cantorian sets are strongly cantorian, and he shows that this is strong. Essentially there are no non-trivial positive results relating different kinds of small sets.

2.3 The Kaye–Specker equiconsistency lemma

The reader is advised to reread section 1.1.2 before proceeding further here. In particular, if Φ is an expression in the language of simple type theory let Φ^+ be the result of raising all type indices in Φ by 1, and let Φ^n be the result of doing this n times. The *ambiguity scheme for* Γ is $\{\Phi \longleftrightarrow \Phi^+ : \Phi$ a closed sentence in $\Gamma\}$, written "*Amb*(Γ)". Full ambiguity is the scheme of all $\Phi \longleftrightarrow \Phi^+$. A "tsau" (*t*ype-*s*hifting *au*tomophism) of a model $M \models$ TST is an isomorphism $M \to M^*$. Thus, if σ is a tsau then $\sigma`x_n$ is an object of type $n+1$ for all objects x_n of type n.[29]

THEOREM 2.3.1 Specker [1962]. *Given a model* M *of* TST *plus full ambiguity, there is* M' *elementarily equivalent to* M *with a tsau* σ.

Proof: This version of Specker's proof is due to Richard Kaye.

Let $M \models \mathrm{TST} + \mathit{Amb}$. We may assume M is countable. Let M'' be an $\aleph_1$-saturated elementary extension on M. We shall construct a countable $M' = \{a_{i,j} : i, j \in \mathbb{N}\} \prec M''$ with a tsau σ as follows.

The construction of the tsau σ and M' is by a modified back-and-forth argument, and we present one step of it. Suppose, inductively, we have tuples $\vec{a}_i = a_{i,1}, \ldots, a_{i,l} \in M'$, each of the same finite length l, $a_{i,j}$ having type i in M', and that M' satisfies the obvious analogous ambiguity scheme with parameters from $\vec{a}_0, \vec{a}_1, \vec{a}_2, \ldots$, that is $\phi(\vec{a}) \longleftrightarrow \phi^+(\vec{a}^*)$, where $a^*_{i,j} = a_{i+1,j}$.

Given an arbitrary $x = a_{i,l+1}$ of type i in M', we show how to construct suitable $a_{0,l+1}, \ldots, a_{i-1,l+1}, a_{i+1,l+1}, a_{i+2,l+1}, \ldots$, so that the ambiguity scheme is true for formulae involving these new parameters. Since x is arbitrary, this will allow us to arrange that $M' = \{a_{i,j} : i, j \in \mathbb{N}\}$ is an *elementary* substructure of M'' in the usual way. At the end of the construction, we will have exhausted all of M', and so the map $\sigma : a_{i,l} \mapsto a_{i+1,l}$ will be the required tsau.

For simplicity let us assume $x = a_{0,l+1}$ is of type 0. We must first find $a_{1,l+1}$ satisfying

$$p(y) = \{\phi^+(\vec{a}^*, y) : M' \models \phi(\vec{a}, x), \phi \in \mathcal{L}_{TST}\}.$$

$p(y)$ is clearly finitely satisfied by applying the ambiguity scheme to conjunctions $(\exists v) \bigwedge \Phi(\vec{a}, v)$ which are true in M' for $v = x$. By saturation $p(y)$ is therefore realized in M' by $a_{1,l+1}$, say.

To find $a_{2,l+1}$ we argue similarly, wishing to realize

$$q(w) = \{\phi^+(\vec{a}^*, a_{1,l+1}, w) : M' \models \phi(\vec{a}, a_{0,l+1}, a_{1,l+1}), \phi \in \mathcal{L}_{TST}\}.$$

This is also finitely satisfied, as can be seen by considering the ambiguity scheme for conjunctions of the form $(\exists v) \bigwedge \Phi(\vec{a}, a_{0,l+1}, v)$ true for $v = a_{1,l+1}$ and by our choice of $a_{1,l+1}$ above.

And so on.

If the original choice of x is not of type 0 we can carry out this construction "downwards" as well as "upwards" in the same way. ■

[29] Strictly these maps are automorphisms only if M is a model of *TNT* rather than TST. In Boffa and Casalegno [1985] these maps are called "S-isomorphisms".

If $M \models$ TST, we can construct a new structure with domain the set of things in M of type 0, equality in the sense of M, and $x \in y$ iff $M \models x \in \sigma`y$. The proof that this resulting structure is a model of *NF* is a simple exercise along the lines of lemma 2.1.5 in section 2.1.2, and will be left as an exercise for the reader.

This tells us that if *TNT* plus complete ambiguity is consistent, then so is *NF*. The converse is even easier, since if M is a model of *NF* we obtain a model of *TNT* plus complete ambiguity by making Z copies of it $(\langle M \times \iota`z : z \in \mathrm{Z}\rangle)$ and saying $\langle x, n\rangle$ is "in" $\langle y, n+1\rangle$ iff $M \models x \in y$.

COROLLARY 2.3.2 Specker's equiconsistency lemma.
NF is equiconsistent with TST *plus full ambiguity.*

See also Crabbé [1975, 1978b], Boffa [1977b], and Dzierzgowski [1993]. Richard Kaye [1991] has recently proved a powerful and subtle generalization.

LEMMA 2.3.3 Kaye. *Suppose that* $M = \langle M_0, M_1, M_2 \ldots\rangle$ *is a structure for the language of* TST *and that* Σ *is the class of formulae of the form* "$\exists\vec{x}\Phi(\vec{x},\vec{y})$" *for* Φ *in some class* Δ *which contains all atomic formulae and is closed under conjunction and substitution of variables and contains* $\psi^+(\vec{y})$ *whenever it contains* $\psi(\vec{x})$. *Suppose further that* $M \models Amb(\Sigma)$. *Then there is a structure for the signature* $\langle \in, =\rangle$ *that satisfies any* σ *of the form* $\forall\vec{y}\Phi(\vec{y})$, *where the result of adding suitable type indices to* Φ *is true in* M *and the* $\mathcal{L}_{TST}$ *formula corresponding to* Φ *is in* Σ.

By compactness *TNT* is equiconsistent with TST, and proofs analogous to the above work as well. It is natural to ask for which Γ the scheme *Amb*(Γ) over *TNT* is consistent. We will briefly treat a few results on this subject, before turning to the subsystems of *NF*, many of which arise from positive results here.

The following version of a result of Grishin [1969] appeared in Boffa and Crabbé [1975]:

PROPOSITION 2.3.4 Grishin. *Amb for formulae containing only two distinct type indices (Amb(2-strat)) is provable in TNT.*

Proof: Since "$x \in y$" is the same as "$\iota`x \subseteq y$", the only things we can say with only two type indices are things expressible in the language of boolean algebra, and among the things *TNT* is going to prove is that each type is an infinite (i.e. of size at least n for each concrete n) boolean algebra. As it happens, the theory of infinite atomic boolean algebras is complete, so that *TNT* actually proves *Amb* for formulæ with only two types. The idea is that every infinite atomic boolean algebra is elementarily equivalent to a countable atomic boolean algebra which is *saturated in the sense of Grishin*, namely one where any element dominating infinitely many atoms is the join of two disjoint elements each dominating infinitely many atoms. It is not hard to show that a countable atomic boolean algebra is saturated in the sense of Grishin iff the quotient algebra modulo the finite sets is atomless. It is a simple matter to show that any two boolean algebras ($\mathcal{B}$ and $\mathcal{C}$) that are saturated in this sense are isomorphic. Map the atoms of $\mathcal{B}$ onto the atoms of $\mathcal{C}$ with g (so that g acts on finite sets in general), and then map the quotients onto each other with h. We have selection functions f and b. A point x in $\mathcal{B}$ belongs to some quotient class $[x]$. Make a note of $x \Delta f`[x]$. $f`[x]$ goes to $b`(h`[x])$, so x goes to $b`(h`[x]) \Delta g`(x \Delta f`[x])$. ■

This seems to be the only interesting Γ for which we know that $TNT \vdash Amb(\Gamma)$. I conjecture that TNT also proves $Amb(\exists_2)$ and even $Amb(\Sigma_1^{Levy})$, but most of the fragments of the forthcoming proofs are not publishable. One exception is lemma 2.4.6 below. In any case, since $Amb(\Sigma_1^{\mathcal{P}})$ is consistent with TNT (this is proposition 2.3.6), it does not much matter whether $TNT \vdash Amb(\Sigma_1^{Levy})$ or not.

REMARK 2.3.5 *$Amb(\Sigma_1^{\mathcal{P}})$ is not provable in TNT.*

Proof: AC is $\Pi_1^{\mathcal{P}}$—"$(\forall x)(\forall y \subseteq \mathcal{P}`x)(y$ a partition of $x \to \exists z \in \mathcal{P}`x$ (z a selection set for y))"—and there is a model of TNT where AC is true at negative types and false at positive types. This is because there are models of TST in which AC holds in precisely the first n types. An ultraproduct gives us a model where AC is true at types with negative indices and false elsewhere. ■

PROPOSITION 2.3.6 *$Amb(\Sigma_1^{\mathcal{P}})$ is consistent with TNT.*

Proof: If $M \models$ TST then $M \hookrightarrow_e^{\mathcal{P}} M^*$ as follows. Let $i_0`x$ be $\iota`x$, and thereafter $i_{n+1}`x_{n+1} = i_n``x_{n+1}$. Evidently $M^*$ is an end-extension of $i``M$ preserving power set. This implies that $\vdash \Phi \to \Phi^+$ for $\Phi \in \Sigma_1^{\mathcal{P}}$. This in turn means that a simple compactness argument gives the consistency of TST $+Amb(\Sigma_1^{\mathcal{P}})$, and thus of $TNT + Amb(\Sigma_1^{\mathcal{P}})$. ■

Specker's reduction provides the setting for one of the most elegant exercises in truth definitions this area has afforded. As I remarked earlier, it is always a good idea to orient oneself in a new set-theoretic environment by trying to derive the paradoxes. We have seen already why the paradoxes (apparently) are not derivable in *NF*, and it is now time to consider some slightly subtler troubles. Why can we not prove in *NF* that every axiom of *NF* is true in *V*? This is worth thinking about in some detail. We will have clauses in our definitions that say, for f an assignment function sending variables (whatever they are) to sets, that for example f *sat* $\ulcorner x_i \in x_j \urcorner \longleftrightarrow f`i \in f`j$. It is easy to see that this is unstratified. The proof that every axiom is true is performed by an induction. Although this does not actually prove that we cannot do anything, the difficulties we have with induction on unstratified formulae should persuade us that enough things go wrong for us not to have to worry about paradox! Of course, if $\in$ is a set E, then this clause above becomes 'f *sat* $\ulcorner x_i \in x_j \urcorner \longleftrightarrow \langle f`i, f`j\rangle \in E$ which *is* stratified. This tells us that if *NF* is consistent, then $\in$ is not a set.[30]

Granted, $\in$ is not a set, but various stratified versions of it are, such as $\{\langle \iota`x, y\rangle : x \in y\}$. Can we perhaps use these to get truth definitions and consistency proofs for subsystems of *NF*? Orey's [1964] answer is "yes!"

Let us work in *NF* (for the moment, we shall add some axioms as the construction goes along) and consider a model M of TST_5 (the '5' is for the sake of argument) constructed as: $\iota^4``V$, $\iota^3``V$, $\iota^2``V$, $\iota``V$, V, with the obvious $\in$ relation. Unfortunately, to make this rigorous we have to label each level differently, since the $\iota^4``V$, $\iota^3``V\ldots$ are not disjoint as presented. What we really have is

$$(\iota^4``V) \times \iota`4, \quad (\iota^3``V) \times \iota`3, \quad (\iota^2``V) \times \iota`2, \quad (\iota``V) \times \iota`1, \quad V \times \iota`0$$

and all the levels are disjoint. The "obvious" $\in$ relation for this model is the slightly cumbersome

$x \in^M y$ iff the second component of x is 4, the second component of y is 3, and $\iota^{-4}`$(first component of $x$) $\in \iota^{-3}`$(first component of y)

with similar clauses for other naturals below 4. This $\in^M$ relation is obviously homogeneous, which is going to avoid the problems we had earlier with unstratified conditions on assignment functions. What we have before us is obviously meant to be a model of TST_5, and if that is all we are trying to prove, the "homogenization" of $\in$ has done the trick. (So for any numeral n, $NF \vdash Con(TSTI_n)$.) How about trying to prove that it is a model of TST_5 plus full ambiguity, which we know will give rise to a consistency proof for *NF*? Well, there is certainly a tsau, so all we have to do is use this tsau (which is not a set, after all) to prove the ambiguity scheme. Let us notate the inverse of this tsau with a '!'. Thus, if x is $\langle \iota`y, n+1\rangle$, $!x$ is $\langle y, n\rangle$.

Now let us write **r** for raise, for the function that sends the Gödel number of Φ to the Gödel number of Φ^+. We must also recall familiar notions from the theory of truth definitions,[31] namely that of an assignment function satisfying a formula, for which we will write "$Sat(f,n)$". Orey has a notation $Caf`f$ intended for use when $f$ is an assignment function. $Caf`f = \{\langle Tn, x\rangle : \langle n, !x\rangle \in f\}$. Then we can prove that

[30] We can also prove this more directly: if $\in$ is a set, then so is $\in \cap (V \times \iota``V)$ which is ι, and this cannot be a set.

[31] Orey recommends Wang [1952b] as "the most enlightening exposition" known to him. It can still be strongly recommended. Wang has applied these techniques to *NF* on his own account in [1953].

$$Sat(g,n) \to Sat(Caf^{-1}`g, \mathbf{r}`T^{-1}`n) \text{ and } Sat(g,n) \to Sat(Caf`g, \mathbf{r}`T`n)$$

which are both stratified. This gives us

$$(\forall n)(n = Tn \to (n \in \mathbf{Tr} \longleftrightarrow \mathbf{r}`n \in \mathbf{Tr}))$$

(where **Tr** is the set of Gödel numbers of true sentences). So, clearly, if n is a cantorian integer and the Gödel number of some expression Φ, then $\#\Phi \longleftrightarrow (\#\Phi)^+$ holds in M. This immediately tells us two things.

REMARK 2.3.7 *For all actually finite k, NF proves the consistency of* TST_k *plus any actually finite subscheme of Amb.*

(It is an open question whether or not we can strengthen remark 2.3.7 to read $NF \vdash (\forall n \in \mathbb{N})(Con(\mathrm{TST}_n))$.) And the following

THEOREM 2.3.8 Orey [1964]. *For all actually finite k*

$$NFC \vdash Con(TSTI_k + Amb)$$

and hence

$$NFC \vdash Con(NF).$$

Proof: (Sketch.)

The first part depends on the equivalence of *NF* and NF_k (for some fixed small k) being provable in the arithmetic of *NF*. We know from Hailperin [1944] (see page 26) that *NF* is finitely axiomatizable, and examination of Hailperin's proof shows that this equivalence can indeed be shown in the arithmetic of *NF* (indeed in much simpler systems). In fact it is sufficient to perform this construction with only five types. The second part similarly depends on the equivalence of $Con(NF)$ and $Con(TSTI + Amb)$ being provable in the arithmetic of *NF*. This is true as well, and perhaps the best place to see this is Crabbé [1975, 1978b]. ■

It is now apparent that, if we had performed this construction in $NFC+\Phi$ where Φ is some n-stratified expression, and had used n types, then, in the model M, $\#\Phi$ would be true as well, so we have actually proved the following.

THEOREM 2.3.9 Orey [1964]. $NFC + \Phi \vdash Con(NF + \Phi)$ *if* Φ *is stratified.*

COROLLARY 2.3.10 *The axiom of counting is not a consequence of any consistent set of stratified sentences.*

We will see an application of this construction in section 2.3.1.

We have seen (remark 2.3.7) how, for any actually finite n, $NF \vdash Con(\mathrm{TST}_n)$. In fact it proves $Con(TSTI_n)$ since the axiom of infinity is (demonstrably) true in any of the models that Orey uses. Let us call these *Orey models.* Thus we can speak of *the nth Orey model.* Can we prove that for all n there is an nth Orey model and so construct within *NF* a compactness argument to show $NF \vdash Con(TSTI)$? In fact it does not seem to be possible to prove that for every n there is an nth Orey model. Suppose we have the nth Orey model. How do we get the $(n+1)$th? All we have to do is add a new bottom type $(\iota^{n+1}\text{“}V \times \{n+1\})$ which we can do uniformly by saying it is $\{\langle \iota\text{‘}x, (n+1)\rangle : \langle x, n\rangle \in$ the bottom type of the previous model$\}$. The problem is that when we come to define the $\in$ relation between the two bottom types in the new model, the $(n+1)$ enters as an exponent of $\bigcup$ and this is illegitimate. Accordingly it is not clear that we can show $NF \vdash Con(TSTI)$ by this means (or indeed by any other).

Construction of models in the spirit of Orey's trick here is possible in typed theories as well. Now of course the set

$$(\iota^4\text{“}V_0) \times \iota\text{‘}4 \;\cup\; (\iota^3\text{“}V_1) \times \iota\text{‘}3 \;\cup\; (\iota^2\text{“}V_2) \times \iota\text{‘}2 \;\cup\; (\iota\text{“}V_3) \times \iota\text{‘}1 \;\cup\; V_4 \times \iota\text{‘}0$$

(where the object-language integers of course all have subscript ‘3’) is a perfectly respectable object of type 4 (if we are using Quine pairs). What the trick achieves is having the membership relation of the model as a *set*, as before, so we can reproduce the Orey construction in *TSTI*. For more on consistency proofs within type theory see McNaughton [1953].

We mentioned earlier (section 2.1.1) Pétry's scheme $(\forall n \in \mathbb{N})(\Phi(n) \longleftrightarrow \Phi(Tn))$ over all stratified Φ. By using Orey's model, Pétry has shown the following.

Remark 2.3.11 *For every standard integer k, there is a stratified formula $A_k(x)$ such that, for any k-stratified formula Φ,*

$$NF + (\forall n \in \mathbb{N})(A_k(n) \longleftrightarrow A_k(Tn)) \;\;\vdash\;\; \Phi \to Con(NF + \Phi).$$

Pétry will publish a proof of this soon.

Before proceeding to consider in detail some of the subsystems of *NF* that these equiconsistency lemmas have shown the consistency of, we should mention some other restrictions of full ambiguity. In Forster [1976] I pointed out that the scheme Amb^n, consisting of $\Phi \longleftrightarrow \Phi^n$, is actually a subscheme of *Amb*, because $\Phi \longleftrightarrow \Phi^n$ is $\Psi \longleftrightarrow \Psi^+$ where Ψ is $(\Phi \longleftrightarrow (\Phi^+ \longleftrightarrow (\Phi^{**} \longleftrightarrow \ldots \Phi^{n-1})))$. I also proved that this scheme suffices to prove the axiom of infinity. Crabbé in [1984] shows that only two instances of Amb^n are needed. Simply let Φ be “$\overline{\overline{V}}$ is $\beth_n\text{‘}0$ and n is even”. No one has yet shown that a model of Amb^n must give rise to a model of *NF*, nor that it need not.

By a suitable variant of lemma 2.3.3 (Kaye's lemma), $Amb(\exists_2)$ (which is conjectured to be provable in *TNT*) would give us the consistency of a $\forall_3$ theory which we expect to turn out to be *NFO*. But, since *NFO* is a subset of NF_3, we already know that it is consistent! Kaye (so far unpublished) has shown that *NF* has a stratified $\forall_5$ axiomatization. It follows from this that $Con(NF) \longleftrightarrow Con(\mathrm{TST} + Amb(\exists_4))$.

2.3.1 NF_3

We have seen that if ϕ is a formula mentioning only two types then $TNT \vdash \phi$ or $TNT \vdash \neg\phi$, so by use of general model-theoretic nonsense in the style of Kaye's lemma (lemma 2.3.3),[32] we infer the consistency of NF_3 (Grishin's result in [1969]).

Some suitable version of Kaye's lemma will prove that if M is a model of TST with bottom type infinite, then the set of 3-stratifiable sentences true in M is a consistent theory, so we have immediately the consistency of many extensions of NF_3. In particular, NF_3 neither proves nor refutes the axiom of infinity. What can we add to NF_3 to get *NF*? This of course has infinitely many answers, but the following two (slight variants of Grishin's result in [1972a] and [1973a]) are attractive and well known. Both have the consequence that $NF = NF_4$.

PROPOSITION 2.3.12 Crabbé [1976]. $NF = NF_3 + (\forall x)(B\text{``}x \textit{ exists})$.

Proof: We want to derive some instance $\forall\vec{x}\,\exists y\forall z(z \in y \longleftrightarrow \Phi(\vec{x}, z))$ of the comprehension scheme. Now suppose 'w' is a variable of minimal type in Φ. That means 'w' only occurs to the *left* of an $\in$. Now '$w \in x$' is equivalent to '$x \in B\text{`}w$', so we can replace all occurrences of things like '$w \in x$' by the corresponding '$x \in w_1$', where 'w_1' is a new variable two types higher than the old 'w', as long as we restrict w_1 to $B\text{``}V$. Thus the desired $\forall\vec{x}\,\exists y\forall z(z \in y \longleftrightarrow \Phi(\vec{x}, z))$ will follow from $\forall X\forall\vec{x}\,\exists y\forall z(z \in y \longleftrightarrow \Phi^X(\vec{x}, z))$, where '$\Phi^X$ is the result of rewriting all occurrences of '$w \in x$' inside 'Φ' in the above style, with X for $B\text{``}V$. At least, it will follow as long as the existence of $B\text{``}V$ is an axiom, for then we just use universal instantiation. Now how many levels does '$\forall X\forall\vec{x}\,\exists y\forall z(z \in y \longleftrightarrow \Phi^X(\vec{x}, z))$' use? Well, as long as the original comprehension axiom used at least four types, this will use one fewer. Therefore we can iterate this procedure to reduce all the comprehension axioms to things using only three types (though possibly many more parameters). ■

PROPOSITION 2.3.13 Boffa [1977a]. $NF = NF_3 +$ *"there exists a set of pairs* $\{\{\iota\text{`}x, y\} : x \in y\}$ *coding the* $\in$ *relation"*.

[32] Kaye's lemma applies only to formulae closed under conjunction.

Proof: The proof of this is quite similar: as before, we suppose 'w' to be a variable of minimal type in Φ. Let E be $\{\{\iota`x, y\} : x \in y \in V\}$ for the nonce. Then 'w' occurs only to the *left* of '$\in$', and we can rewrite $\ulcorner w \in x_i \urcorner$ with $\ulcorner w_1 \in \iota``V \wedge \{w_1, x_i\} \in E \urcorner$, where '$w_1$' is a new variable (this time *one* type higher than 'w'). $=$ gives no problems; so, as before, we have used one fewer type (as long as we were using at least four when we started) but more parameters. It is easy to check that the existence of $\iota``V$ is an axiom of NF_3. ∎

The collection $\{\{\iota`x, y\} : x \in y\}$ is of course a set in *NF*, for it is the extension of a stratified set abstract. For the same reason, corresponding set abstracts exist at each type in TST. We have seen in the last section how the sethood of this object enables us to construct truth definitions for stratified formulae using boundedly many types in *NF*. The fact that typed versions of this set also exist in TST means that if $M \models$ TST then for each initial segment M' of M, M contains truth definitions for the language of M'.

Boffa in [1975b] explores what happens when one tries to modify Orey's construction from the preceding section with NF_3 in mind. One can work in *NF* as before, up to the point where we need to prove that *Amb* holds in M. If we only have three levels, then all we have to prove is that $M \models$ *Amb*(2-*strat*), since that is all the ambiguity M can express, and *this* much ambiguity is actually provable, as we have just seen. So *NF* proves $Con(NF_3)$. Hardly surprising, one might think, but (just as in the previous construction) if we started in a theory $NF + \Phi$, where Φ is some 3-stratified expression, then we could certainly show that $\#\Phi$, the appropriate typed version of Φ, is true in M, so what this method really tells us is

PROPOSITION 2.3.14 Boffa [1975b]. *If Φ is 3-stratifiable, then $NF + \Phi \vdash Con(NF_3 + \Phi)$.*

This has many corollaries: for example, a simple application of compactness as follows.

COROLLARY 2.3.15 *NF has no (consistent) axiomatization consisting entirely of 3-stratified sentences.*

There is a 3-stratified version of the axiom of infinity. We may define x to be finite iff it belongs to all sets which extend $\iota``V$ and are closed under $\cup$. This is 3-stratified, so the assertion that there is a set that is not finite is 3-stratified. Therefore, since the axiom of infinity is a theorem of *NF* we have the following.

COROLLARY 2.3.16 $NF \vdash Con(NF_3 + AxInf)$.

2.3.2 *NFU*

The best references for the study of *NFU* are Crabbé [1992] and the publications of Randall Holmes.

It seems to be fairly widely known that *NFU* is consistent (Jensen [1969]) and does not prove the axiom of infinity. Even apart from the question "Why?", there is also the question "What goes wrong in the proof of the axiom of infinity in *NF*? I don't see any use of extensionality!". The consistency of $NFU + \neg AxInf$ is one of those things that upsets readers by giving them the sudden feeling that they have not, after all, understood something that had seemed clear.

The answer is that the proof of the axiom of infinity in *NF* depends on the fact that T commutes with exponentiation and that $V = \mathcal{P}`V$. In *NFU* the *urelemente* complicate the definition of power set because we have to choose *one urelement* to be $\Lambda$. $V$ contains *all urelemente* whereas $\mathcal{P}`V$ contains just the one we have chosen to be Λ. Thus it might turn out that V is bigger than $\mathcal{P}`V$. So it is no longer the case that $2^{T\overline{\overline{V}}} = \overline{\overline{V}}$, and the proof of the axiom of infinity collapses. Anything we can add to *NFU* to restore this state of affairs will restore the proof of the axiom of infinity. If there are sufficiently few *urelemente* then the universe can be the same size as the set of non-empty sets and the proof will go through. This remark is worth minuting:

REMARK 2.3.17 *In any model of NFU that is not, and does not naturally give rise to, a model of NF, the set of atoms is so large that its complement is strictly smaller than the universe.*

Proof: If π is a set that maps $V \longleftrightarrow \mathcal{P}`V$, then we get a model of *NF* with domain $V$ and membership relation $x \in_\pi y$ iff $x \in \pi`y$. For more detail on this see Boffa [1973]. ■

THEOREM 2.3.18 (Holmes) (*NFU*+ *AC*): *For each concrete* n, $\beth_n\, \overline{\overline{\mathcal{P}`V}} < \overline{\overline{V}}$

Proof: Suppose that the theorem is false. Then there is a concrete n such that $\beth_n\, \overline{\overline{\mathcal{P}`V}}$ does not exist. Let $n$ be the smallest such. Observe that $\beth_{n+1}\, \overline{\overline{\mathcal{P}`V}}$ does not exist, and that $\beth_n\, \overline{\overline{\mathcal{P}`\iota``V}}$ does exist.

Let m be the smallest cardinal such that $\beth_i(m)$ does not exist for some i. Let $j+1$ be the smallest such i. Now look at the sequence of iterated images of Tm under exponentiation. The $Tj+1$st element of this sequence exists and is greater than $T\,\overline{\overline{V}} = \overline{\overline{\iota``V}}$, so it has no more than n iterated images under exponentiation; between 1 and $n+1$ new terms are added to the sequence. Thus, the number of terms in the sequence for Tm is finite and differs from the number of terms in the sequence for m mod $n+2$

(say); recall that n is standard, so m is different from Tm. Thus $m < Tm$ (by minimality of m). But then $T^{-1}m < m$, and $T^{-1}m$ is easily seen to have between 1 and $n+1$ fewer terms in its sequence of iterated images under exponentiation than m, violating minimality of m. ■

There are several proofs of the consistency of *NFU*. The prettiest is undoubtedly that due to Boffa [1977b], which depends on Ramsey's theorem.

THEOREM 2.3.19 (Jensen [1969]). *NFU is consistent.*

Proof: (Boffa.)

Let $M = \langle M_i : i \in \mathbb{N}\rangle$ be a model of TST. For $I \subseteq \mathbb{N}$, let the *extracted model* M_I be $\langle M_i : i \in I\rangle$ with a new $\in$ relation. We say $x_{i_n} \in y_{i_{n+1}}$ iff y is a set of singletons$^{i_{n+1}-i_n-1}$ (otherwise y is an *urelement*)[33] and $\iota^{i_{n+1}-i_n-1}$'x is a member of y in the sense of M. We check that M_I is a model of *TSTU*, that is TST with *urelemente.*

Now let $M \models$ *TSTU*, and let Φ be an arbitrary expression in $\mathcal{L}_{\mathrm{TST}}$. Φ speaks of, say, five types. Let us partition $[\mathbb{N}]^5$. Let $I = \{i_1, \ldots, i_5\}$ and send $\{i_1, \ldots, i_5\}$ to 1 if $M_I \models \Phi$ and to 0 otherwise. We now invoke Ramsey's theorem to find an infinite $J \subseteq \mathbb{N}$ homogeneous for this partition and consider M_J. By homogeneity, either every model extracted from $M_J \models \Phi$ or every model extracted from $M_J \models \neg\Phi$, so certainly every model extracted from M_J satisfies $\Phi \longleftrightarrow \Phi^+$. We now repeat the process for a different Φ. This shows that for any finite collection of formulae $\langle \Phi_i : i \in I\rangle$, we can find a model of $TSTU + \bigwedge_{i \in I} \Phi_i \longleftrightarrow \Phi_i^+$. By compactness, we have a model of *TNTU* plus complete ambiguity, and thus of *NF*U by Specker's lemma. ■

This proof has the advantage of showing that *NFU* is no stronger than TST without the axiom of infinity. It is possible to produce models of *NFU* directly and in bulk by cruder techniques.

PROPOSITION 2.3.20 *If $\langle V, \in\rangle$ is a model of Z (Zermelo set theory) with an (external) automorphism σ and an ordinal κ such that $\kappa > \sigma$'κ, then V_κ is the domain of a model $\mathcal{M}$ of NFU whose membership relation ($\in_{\mathcal{M}}$) is*

$$\{\langle x, y\rangle : \sigma\text{`}x \in y \wedge \rho\text{`}y \leq \sigma\text{`}\kappa\}.$$

Proof: The true sets of $\mathcal{M}$ are the sets of rank $\leq \sigma$'κ. Anything of rank $> \sigma$'κ will be an *urelement*, and $V_{\sigma\text{`}\kappa}$ will be the universal set of $\mathcal{M}$.

To show that $\mathcal{M}$ satisfies all the comprehension axioms of *NFU*, it is sufficient to show that all of them translate into expressions without any occurrences of 'σ' in them, for then we can appeal to the comprehension scheme available internally in the model of Z. We want $\mathcal{M} \models \{z : \Phi(z, \vec{a})\}$

[33] The superscript is the number of times that ι is to be iterated.

exists. We write out $\Phi(z, \vec{a})$ in full, with '$u \in_{\mathcal{M}} v$' replaced by '$\sigma`u \in v \wedge \rho`v \leq \sigma`\kappa$'. Now we want to know that the set of all things of rank $< \kappa$ satisfying this rewritten version of $\Phi$ is indeed a set and is of rank at most $\sigma`\kappa$.

We only have to prove this in the case where Φ is stratified. Fix some stratification of Φ.[34] Any variable in a stratified formula has some type in a stratification. Since σ is an automorphism, we have

$$(\forall x)(\forall y)(\sigma`x \in y \longleftrightarrow \sigma^{n+1}`x \in \sigma^{n}`y)$$

for any n. Indeed—which is more to the point—we also have

$$(\forall x)(\forall y)[(\sigma`x \in y \wedge \rho`y \leq \sigma`\kappa) \longleftrightarrow (\sigma^{n+1}`x \in \sigma^{n}`y \wedge \rho`(\sigma^{n}`y) \leq \sigma^{n+1}`\kappa)]$$

and by substitutivity of the biconditional we can ensure that every occurrence of any particular variable has the same prefix (σ^n for some n) in front of it, where the 'n' depends only on the type of that variable according to the stratification. Now we find that the comprehension axiom has quantified variables x such that every occurrence of 'x' has the same prefix σ^n. Therefore we can rewrite $\forall x \ldots$ as $\forall x \in \sigma^n``V_\kappa \ldots$, and delete the prefix. But $\sigma^n``V_\kappa$ is an object inside V and therefore is allowed to appear as a parameter in a separation axiom in the model of Zermelo that we are considering. The clause "$\rho`y \leq \sigma`\kappa$" in the definition of the $\in$ relation of the model gives rise to occurrences of '$\sigma^{n}`\kappa$' for various n but all these, too, are allowed to appear as parameters. Once we do this relettering for all variables, we have reduced the *NFU* comprehension axiom to a separation axiom containing some extra parameters. Of course if Φ had bound variables of type more than (1 + type of 'z') then in the process of applying sufficiently many 'σ' to our variables to ensure that the prefix is constant for any given variable, we may find that we have replaced $V_{\sigma`\kappa}$ by $\sigma^{n}`V_{\sigma`\kappa}$; but, since $\sigma$ is an automorphism, the fact that the right subset of $\sigma^{n}`V_{\sigma`\kappa}$ exists is enough for the right subset of $V_{\sigma`\kappa}$ to exist. Finally we have to check that the collection whose existence is assured by internal separation is also coded in $\mathcal{M}$. The collection (let us call it 'PHI') contains things potentially of all ranks below κ but nothing of rank as great as κ and therefore is of rank at most κ. If b was some object such that $\Phi(b, \vec{a})$ we will have $b \in$ PHI, and therefore $\sigma`b \in \sigma`$PHI. So $\sigma`$PHI is the object which is—in the sense of $\mathcal{M}$—the set of all things that are $\Phi$. Fortunately $\sigma`$PHI is of rank $\sigma`\kappa$ at most and is therefore not an *urelement*! ■

[34]The problem of getting rid of all occurrences of 'σ' is really the same problem as we will see in the introductory sections of chapter 3, and the reader should consult that before reading further here.

Models of Z of the kind needed in proposition 2.3.20 can be obtained by use of the Ehrenfeucht–Mostowski theorem (for example). There are two points that should be noted here. (i) Successfully verifying *NFU*-style comprehension in the new model depends on being able to eliminate occurrences of 'σ' in favour of parameters. This depends crucially on the formulae in question being stratified. (ii) Once we have eliminated the occurrences of 'σ' in favour of parameters, the comprehension axiom that we need to verify in the old model has the feature that all its bound variables are restricted to parameters. This means that we need only Δ_0-separation to hold in the original model, and that accordingly systems much weaker than Z will suffice.

The proof is implicit in Boffa [1988], and was noticed also by Kaye.

In the construction of proposition 2.3.20 anything of rank $> \sigma`\kappa$ will be an *urelement.* Thus the closer $\sigma`\kappa$ is to κ the "fewer" (in some sense) *urelemente* the resulting model will have. We assumed the existence of only one external automorphism σ such that for some κ, $\kappa > \sigma`\kappa$, but it is not unreasonable to contemplate the existence of a family $\{\sigma_i : i \in I\}$ of external automorphisms all pressing $\kappa$ down such that the sup of $\{\sigma_i`\kappa : i \in I\}$ is actually κ itself. One might hope that the resulting models would fall naturally into a commutative diagram with embeddings which would give us a limit of some sort which would be a model of *NF*. This has never been systematically investigated: it is clear neither what the consistency strength is of the existence of such a family of automorphisms, nor what should be the embedding relation for the resulting family of models.

2.3.2.1 *How different can atoms be?* We remarked earlier (remark 2.3.17) that in any model of *NFU* that does not immediately give rise to a model of *NF* the set of atoms is so large that its complement is strictly smaller than the universe. If a theory requires there to be lots of atoms that are *distinct*, one might expect that it should also make them disti*nguishable* in some way. This suggests that the assertion "*NFU* has a model $\mathcal{M}$ such that $Aut(\mathcal{M})$ acts transitively on tuples of atoms from $\mathcal{M}$" is strong. It is certainly true if (and perhaps *only* if!) *NF* is consistent. How strong *is* this assertion? Established techniques seem to be able to give us only a partial answer. Clearly the question can be related to questions about how many (or how few) types can be realized by the atoms in a model of *NFU*.

THEOREM 2.3.21 *In all models of NFU all tuples of distinct atoms realize the same weakly stratified types.*

We shall obtain this as a consequence of the following lemma.

LEMMA 2.3.22 *If σ is a permutation that moves atoms only, and Φ is a weakly stratified formula, then*

$$\Phi(\vec{a}) \longleftrightarrow \Phi(\sigma\text{`}\vec{a})$$

where the $\vec{a}$ range over atoms.

Proof: We know from lemma 2.1.5 that if Φ is stratified and s_i is the type of 'a_i' in Φ then

$$\Phi(\vec{a}) \longleftrightarrow \Phi(\ldots(j^{s_i}\text{`}\sigma)\text{`}a_i\ldots).$$

The permutations we are going to restrict ourselves to are special so we can get a bit more. When a is an atom we have a prima facie problem with expressions like

$$(j^3\text{`}\sigma)\text{`}a$$

because all we know about this set abstract is that it is empty. Which empty set is it? It turns out that we can deal with this by stipulation. For each $n \geq 0$ we stipulate that $(j^{n+1}\text{`}\sigma)\text{`}x$ is to be
1. $\{(j^n\text{`}\sigma)\text{`}y : y \in x\}$ if x is non-empty and
2. $\sigma\text{`}x$ otherwise.

This enables us to show that

$$(\forall x)(\forall y)(x \in y \longleftrightarrow ((j^n\text{`}\sigma)\text{`}x \in (j^{n+1}\text{`}\sigma)\text{`}y)).$$

Proof: If y is an atom then $(j^{n+1}\text{`}\sigma)\text{`}y$ is $\sigma\text{`}y$ which is also an atom so both halves of the biconditional are false, so the biconditional itself is true. So suppose $y$ is non-empty. Then by clause 1 above $(j^{n+1}\text{`}\sigma)\text{`}y = \{(j^n\text{`}\sigma)\text{`}w : w \in y\}$.

left $\rightarrow$ right

If $x \in y$ it follows from this definition that $(j^n\text{`}\sigma)\text{`}x \in (j^{n+1}\text{`}\sigma)\text{`}y$ as desired.

right $\rightarrow$ left

Suppose $(j^n\text{`}\sigma)\text{`}x \in (j^{n+1}\text{`}\sigma)\text{`}y$. This gives $(j^n\text{`}\sigma)\text{`}x \in (j^{n+1}\text{`}\sigma)\text{`}y = \{(j^n\text{`}\sigma)\text{`}w : w \in y\}$, whence $(j^n\text{`}\sigma)\text{`}x \in \{(j^n\text{`}\sigma)\text{`}w : w \in y\}$. To get $x \in y$ from this we need to know that $j^n\text{`}\sigma$ is a permutation of V. Normally of course this is routine but this time there are atoms about so we have to be careful.

Clearly $j^n\text{`}\sigma$ is defined everywhere. We have to show it is onto. Every atom $a$ is $j^n\text{`}\sigma$ of something, since it is $j^n\text{`}\sigma$ of $\sigma^{-1}\text{`}a$. For non-atoms $j^n\text{`}\sigma$ has its usual meaning. ■

Now we return to our Φ where the only free variable places are occupied by variables ranging over atoms, and where Φ is weakly stratified. That is to say, we can give type assignments consistently to all the bound variables. Thus a bound variable 'x_i' receives type τ_i in some stratification assignment, and all occurrences of it will be replaced by '$(j^{\tau_i}\text{`}\sigma)\text{`}x_i$'. The free variables are atoms and will not have all their occurrences replaced in this uniform way, but since $(j^n`\sigma)`a$ is always $\sigma`a$ irrespective of n this does not matter.

This proves $\Phi(\vec{a}) \longleftrightarrow \Phi(\sigma`\vec{a})$ and thus lemma 2.3.22. Now σ was an *arbitrary* permutation of the set of atoms and since the symmetric group on the set of atoms clearly acts transitively on n-tuples of (distinct) atoms we have at least shown that all n-tuples of (distinct) atoms realize the same (weakly stratified) types, which proves theorem 2.3.21. ■

Notice that in the model obtained in the proof of proposition 2.3.20 the atoms are grossly distinguished by their ranks, and it seems impossible for a model arising in this way to have indiscernible atoms. To see how hard this is, consider the binary relation on atoms: $\forall x(x \in x \rightarrow (a \in x \longleftrightarrow b \in x))$.

For an extended application of (an extension of) *NFU* see Feferman [1972].

2.3.3 *Lake's model*

We start with a countable model of *NFU* and consider a sequence of equivalence relations $\sim_\alpha$, where $\sim_0$ identifies all empty sets but nothing else, and $\sim_{\alpha+1}$ is $\sim_\alpha^+$, in which $+$ is Hinnion's $+$ operation from section 1.2.1.1. Evidently at some countable stage, we reach a fixed point (a contraction) and the quotient will be a model for extensionality. The quotient is non-trivial, since V and Λ can never become identified! It turns out (though Lake [1974] proves P1 only) that it is a model of all the Hailperin axioms from section 2.1 except possibly P6.

This can also be conducted by considering substructures instead of quotients. We start with a countable model M_0. $M_{\alpha+1}$ is derived from M_α as a selection set from the set of equivalence classes $x \sim y$ iff $x \cap M = y \cap M$. Take intersections at limits. Thus, for $\alpha > \beta$, M_β is a substructure of M_α, and there is a homomorphism from M_α onto M_β.

2.3.4 KF

We have just learned that $Amb(\Sigma_1^{\mathcal{P}})$ is consistent with *TNT*, so by lemma 2.3.3 (Kaye's lemma) we have the consistency of the theory below, which we call KF.

1. Extensionality
2. Pair set
3. Power set
4. Sumset
5. Stratified $\Delta_0^{\mathcal{P}}$ *separation.*

This theory is identical with the "Maclane set theory" of Mathias [199?] except for the restriction to stratified formulae in 5.

We reproduce here without proof some of the results in Forster and Kaye [1991].

REMARK 2.3.23 KF *is equivalent to* 1–4 *above plus the axiom scheme*

$$\forall x \langle x, \mathcal{P}`x, \mathcal{P}^{2}`x, \ldots, \mathcal{P}^{n}`x \rangle \models \sigma$$

for each axiom σ *of* TST, *with* n *depending on* σ.

From the point of view of the present essay, the importance of KF is that it is both the strongest natural set theory obviously true in the well-founded sets of any model of *NF* and the weakest natural set theory compatible with the axiom of foundation such that, if you add $(\exists x \forall y)(y \in x)$ to it, you get *NF*. Clearly KF $\subseteq$ *NF* and KF $\subseteq Z$. It is obvious that, if $M \models NF$, then $WF^M \models$ KF plus the axiom of foundation, and it is the strongest theory that obviously satisfies this. As Kaye points out, since *NF* and *ZF* both prove stratified collection, so does their intersection, but $Th(WF)$ (the set of sentences true in *WF*) presumably does not, so we cannot expect to be able to replace KF by $ZF \cap NF$[35] in this assertion. *KFI* is weaker than Z, *NF*, and quite possibly even $Z \cap NF$. To see this, suppose $Con(KFI) \rightarrow Con(Z \cap NF)$. Then $Con(KFI) \rightarrow Con(Z) \vee Con(NF)$. Should it turn out that $Con(NF) \rightarrow Con(Z)$, we would infer $Con(KFI) \rightarrow Con(Z) \vee Con(Z)$ and $Con(KFI) \rightarrow Con(Z)$. But it is an old result of Kemeny that $Z \vdash Con(TSTI)$; whence $Z \vdash Con(KFI)$, which would give us a contradiction. Unfortunately, no proof of the assertion that $Z \vdash Con(KFI)$ has been published (so far as I am aware, though it is reported in Kemeny [1950] and alluded to by Quine [1967]) and it has turned out to be much harder to prove $Con(NF) \rightarrow Con(Z)$ than I had originally supposed. Indeed it is still open.

The following remarks give some clue as to what might happen if we add other "universal" objects (set of all cardinals, ordinals, ...) to KF. An observation very like the following was made by Kaye.

REMARK 2.3.24 *If* M *is a model of* KF *and* $\bigcup x \subseteq x \in M$, *then*[36] $\bigcup \langle\langle x \rangle\rangle \subseteq_e^{\mathcal{P}} M$ *and is also a model of* KF.

The only subtlety in this proof is that we need x to be transitive in order for this model to satisfy the axiom of sumset. We can rely on $\Delta_0^{\mathcal{P}}$ *separation* in M to give us sumsets of anything that appears in $\mathcal{P}^{n}`x$ with $n \geq 1$, but this is not enough to give us sumsets of members of $x$ unless $x \subseteq \mathcal{P}`x$.

LEMMA 2.3.25 The bounding lemma. *If* $\text{KF} + TC \vdash \forall \vec{x}\, \exists \vec{y}\, \phi$ *with* $\phi \in \Delta_0^{\mathcal{P}}$ *then, for some* n, $\text{KF} + TC \vdash \forall \vec{x}\, \exists \vec{y} \in \mathcal{P}^{n}`TC`\{x_0, \ldots, x_i\} \phi$.

[35] This theory is the intersection of *ZF* and *NF*, understood as the sets of their theorems.

[36] By this we of course mean $(x \cup \mathcal{P}`x \cup \mathcal{P}^{2}`x \ldots) \cap M$.

Proof: Suppose for all words $t(\ )$ in the alphabet with $\bigcup$, TC, $\mathcal{P}$ we have, for some $\phi \in \Delta_0^{\mathcal{P}}$, that

$$\mathrm{KF} + TC \not\vdash (\forall \vec{x})(\exists \vec{y} \in t(\vec{x}))(\phi(\vec{x}, \vec{y})).$$

Now consider the set of sentences $\exists \vec{x}(\forall \vec{y} \in t(\vec{x}))\neg\phi(\vec{x}, \vec{y})$ as t varies. By assumption, they are all individually consistent, and it is easy to see that this set is closed under conjunction. We now expand the language by adding a list of constant $\vec{a}$ and a new axiom

$$(\forall \vec{y} \in t(\vec{a}))\neg\phi(\vec{a}, \vec{y})$$

for each word $t(\ )$ as above. By compactness this theory is consistent and has a model M. Now consider $\bigcup \langle\langle TC\text{`}\vec{a}\rangle\rangle \cap M$ as above. This is evidently a model of KF + $(\exists x)(\forall \vec{y})(\neg\phi(x, \vec{y}))$. ■

Now we can consider applications of the bounding lemma to natural $\Pi_2^{\mathcal{P}}$ formulae connected to the existence of a universal set.[37] Set theories with a universal set may furnish us with objects universal in other senses: a set containing sets of all sizes, or well-orderings of all order-types, and the assertions that there are no such universal objects are $\Pi_2^{\mathcal{P}}$ and will be reduced by lemma 2.3.25, the bounding lemma, to something tractable in well-founded set theory. Such an analysis may well reveal dramatic differences in the strengths of these universality assertions. For example, it is not immediately obvious that the existence of a set containing sets of all sizes, or well-orderings of all order-types, implies the existence of a universal set. Driving such a wedge between Cantor's paradox and the Burali-Forti paradox could be one of the most radical developments of our understanding of the paradoxes since the incompleteness theorem.

Accordingly let us consider the expression $\exists NO$:

$$\exists X(\forall \text{well-orderings } y)(\exists x \in X)(y \simeq x).$$

Since there is no unstratified *separation* in KF, we cannot prove that the order-type of the set X' of well-orderings of subsets of X is too long to be embedded in X, just as we cannot do this in *NF*. Suppose we have a model M of KF containing a transitive x such that later members of the sequence x, $\mathcal{P}\text{`}x$, $\mathcal{P}^2\text{`}x$, ... do *not* contain longer well-orderings than earlier members. Evidently, by the remark preceding lemma 2.3.25, the bounding lemma, $\bigcup \langle\langle x\rangle\rangle^M$ will be a model of KF in which there is a set (the set of isomorphism classes of well-orderings of subsets of x) witnessing the

[37] It is worth noting that there is a use of compactness here which is not obviously removable. This may turn out to be significant given that there are at least some senses of "ω-standard" under which *NF* has no ω-standard model (see Rosser and Wang [1950]).

existential quantifier in $\exists NO$. To arrange for this it will turn out that it will suffice to prove the consistency of the two schemes below. Find an α so that

$$\alpha \leq 2^{T\alpha}$$

and

$$A_n: \aleph(\alpha) = \aleph(2^{T\alpha}) = \aleph(2^{2^{T^2\alpha}}) = \ldots = \aleph(\beth_n`(T^{n}`\alpha)).$$

If we can find such an α then, for any x such that $\overline{\overline{x}} = \alpha$, there is a map $x \hookrightarrow \mathcal{P}`x$. The idea is that this can be extended without undue sweat and tears to a setlike permutation τ of the universe so that x is a transitive set in the permutation model M^τ (see chapter 3). We then consider $\bigcup \langle\langle x \rangle\rangle^{M^\tau}$, and this will be the desired model of $\exists NO$.

The scheme A_n (with $\alpha \leq 2^{T\alpha}$) is clearly incompatible with α being a natural number, and slightly less clearly incompatible with α being an aleph. At this stage there seems no reason to believe that KF + $\exists NO$ is *NF* or even equiconsistent with it. However, it may well be equiconsistent with *NF*, for one would expect that any consistency proof of *Z* relative to *NF* along the lines that we will discuss in section 2.5 would work also in KF + $\exists NO$. Accordingly we will expect that KF + $\exists NO$ to be significantly stronger than plain *KF*I (which we have seen is provably consistent in *Z*). Another feature of interest is that it would be the first example of a non-trivial consistent set theory (with extensionality) and a (representative) set of all ordinals. Not only that, *but there is at this stage no reason to suppose that it is incompatible with the axiom of foundation.* These two possibilities provide us with the best possible incentive to develop a notion of a constructible model for *KF*I and learn how to do forcing in *KF*I so that we can arrange the failure of *AC* that $\exists NO$ needs.

THEOREM 2.3.26 *There is a finite collection of stratified (but not homogeneous)* Δ_0^{Levy} *with the property that any model of power set and sumset that is closed under all of them is also a model of stratified* Δ_0^{Levy} *comprehension (and conversely).*

Our n-tuples will be Hailperin n-tuples. Hailperin pairs are Wiener–Kuratowski pairs, and the Hailperin $n+1$-tuple $\langle x_1 \ldots x_{n+1}\rangle$ is the pair $\langle \iota^{2(n-1)}`x_1, \langle x_2 \ldots x_{n+1}\rangle\rangle$. The extra iotas are put in to ensure that all the free variables have the same type. (There is a cost attached to this, as we will see below.) Logicians are nowadays much more sensitive to the possibility of their actions being dependent on particular implementations. All we are assuming about the Wiener–Kuratowski implementation is that "$x = \langle y, z\rangle$" is stratified with 'x' being given a type index two greater than the index assigned to 'y' and 'z'.

The reason why we do not use Wiener–Kuratowski ordered n-tuples for $n > 2$ is that in "$x = \langle x_1, \ldots x_n\rangle$", the various '$x_i$' all receive different types in stratification assignments. This makes it extremely difficult to manipulate fomulae in such a way that (once we have interpreted the theory of ordered pairs in the theory of $\in$) the set of stratified formulae have the right closure properties. Hailperin ordered n-tuples do at least have the feature that in "$x = \langle x_1, \ldots x_n\rangle$", the various '$x_i$' all receive the *same* type in stratification assignments. However, this move creates problems of its own. In the unstratified case we use the usual recursive definition of the $(n+1)$-tuple $\langle x_1 \ldots x_{n+1}\rangle$ as $\langle x_1, \langle x_2 \ldots x_{n+1}\rangle\rangle$. We can then show that there is a small finite number of operations on tuples which enable us to permute the order of components of n-tuples at will. This is of course very useful. The problem with Hailperin n-tuples arises because the operation sending x to $\{\langle\langle u, v\rangle, w\rangle : \langle u, \langle v, w\rangle\rangle \in x\}$ (which is one of the small finite number of operations mentioned above) is not stratified and we cannot use it here.

There are various things we could be trying to prove about stratified Δ_0 separation. We might be trying to show that $\{\langle\vec{x}\rangle \in B : \phi(\vec{x}, \vec{A})\}$ is a set whenever $\phi \in \Delta_0$; we might be trying to prove the existence of $f`x$ for all $x$ and for $f$ such that '$y = f`x$' $\in \Delta_0$. The difference is that the first does not give us the existence of $\iota``x$ but only $\iota``x \cap A$. The slightly annoying feature is that in the finitization below (of the first version of the problem) we use the existence of $\iota``x$ which we seem to get only in the second version. If we can take $A = \mathcal{P}`x$ then we can indeed get $\iota``x$, and can cease to worry about these subtleties. So what we really want is a list of operations s.t. a set closed under pairing and power set and sumset is a model of $str(\Delta_0)$ comprehension iff it is closed under all these operations.

We want to show that

$$\{x : \Phi\} \cap A$$

is a set for any A and any $\Phi \in str(\Delta_0)$. The variables that appear to the left of the colon will be 'x's. There may be parameters inside Φ and they will be 'A's. Subscripts will be provided.

We start by putting Φ into a form with all the quantifiers (which in any case are restricted) out the front and with a matrix that is a disjunction of conjunctions of atomics and negatomics. This will be called *normal form.* We will need to know that every formula can be put into normal form, *and that we do not need any specific set-theoretic assumptions to do this.* This is because we will have to rearrange some formulae into normal form during the proof by induction.

We will need to consider terms like $\{\langle\vec{x}\rangle : \Phi(\vec{x})\}$ for Φ stratified and Δ_0. Although it will suffice to prove the sethood of such things where the tuples are of length 1 (this is because $\{\vec{x} : \Phi(\vec{x})\}$ is $\{y : \Phi^*(y)\}$ for some stratified Δ_0 formula Φ^*; let us write

$$\{\langle \vec{\mathbf{x}} \rangle : \Phi(\vec{x}, \vec{y})\}$$

to mean that the set of tuples we are considering has had its entries prefixed by enough ιs to make $\{\vec{x}{:}\Phi(\vec{x})\}$ into a stratified term) we will prove it for n-tuples of arbitrary length because that is the way the induction works. Of course there are infinitely many ways of doing this, for one can always add one more ι to each coordinate and the result is still a stratified term. Let us call members of this infinite family *versions*.

What we are now going to prove by induction is that

THEOREM 2.3.27 *For any Φ in normal form, any tuple $\vec{x}$, and any parameter A, cofinitely many versions of*

$$\{\langle \vec{\mathbf{x}} \rangle : \Phi(\vec{x})\} \cap A$$

are sets.

Thus, for sufficiently large n, we will prove the existence of $\{\iota^{n}``x : \Phi(x)\} \cap A$ and use $\bigcup$ to get $\{x : \Phi(x)\}$.

We will perform this induction backwards in order to find out which primitive operations we shall need.

Suppose we want infinitely many versions of

$$\{\langle \vec{\mathbf{x}} \rangle : (\exists y \in A_1)\Phi(\vec{x})\} \cap A_2.$$

The obvious thing is to consider something like

$$\{\langle \mathbf{y}\vec{\mathbf{x}} \rangle : (y \in A_1) \wedge \Phi(\vec{x})\} \cap (A_1 \times A_2)$$

and do some kind of projection.

Now any version of

$$\langle \mathbf{y}\vec{\mathbf{x}} \rangle$$

is

$$\langle \iota^{n}`y, \langle \vec{\mathbf{x}} \rangle \rangle$$

for some n (dependent on the version of $\langle \vec{\mathbf{x}} \rangle$) by definition of Hailperin tuples,[38] so what we are really after is

[38]One might think at first that this n is precisely $2k-2$ where k is the length of the tuple $\vec{x}$. It's a bit more complicated than that, for we may have had to put some iotas in front of y to ensure that all the entries in the ordered $(k+1)$-tuple are the same type.

$$\{\langle \iota^n{}^{\text{`}}y, \langle \vec{\mathbf{x}}\rangle\rangle : [(y \in A_1) \wedge \Phi(\vec{x})]\} \cap (\iota^n{}^{\text{``}}A_1 \times A_2)$$

for some n. Let us call this A_3. Cofinally many versions of A_3 will be sets by induction hypothesis, since the matrix of the set abstract (within square brackets) contains one fewer quantifier than before. So if we have an operation $proj(A_1, A_2)$ defined as

$$\{z : \exists y \in A_1 \langle y, z\rangle \in A_2\}$$

then the set we originally wanted is $proj(\iota^n{}^{\text{``}}A_1, A_3)$.

That tells us that so far we need $A_1 \cap A_2$, $A_1 \times A_2$, $proj(A_1, A_2)$, and $\iota^{\text{``}}A$.

It is this inductive clause (and the next) that require us to prove the complicated version of the theorem that is claimed (with the apparatus of *versions*). The point is that if the variable 'y' is of much higher type than the $\vec{x}$s in Φ, then in order to get

$$\{\langle \iota^n{}^{\text{`}}y, \langle \vec{\mathbf{x}}\rangle\rangle : [(y \in A_1) \wedge \Phi(\vec{x})]\} \cap (\iota^n{}^{\text{``}}A_1 \times A_2)$$

to be a stratified term we might have to prefix the $\vec{x}$s with lots of ιs, and thereby lose a few versions.

There is another kind of restricted existential quantifier. The variable might be restricted not to an A, but to an x, so we are looking at

$$\{\langle \vec{\mathbf{x}}\rangle : (\exists y \in x_i)\Phi(\vec{x})\} \cap A.$$

Similarities with the preceding case suggest we should be looking at something like

$$\{\langle \mathbf{y}\vec{\mathbf{x}}\rangle : y \in x_i \wedge \Phi(\vec{x})\} \cap B \times A$$

for some suitable B. (Of course the formula to the right of the colon can be put into normal form without increasing the number of quantifiers.) As before, this is just

$$\{\langle \iota^n{}^{\text{``}}y, \langle \vec{\mathbf{x}}\rangle\rangle : y \in x_i \wedge \Phi(\vec{x})\} \cap B \times A$$

for some n. Let us abbreviate this to 'X'. But what is this B? We know that the witness y we are after is a member of the ith coordinate of some member of A. We had better have a notation for the set of ith coordinates of members of A. Let us write this $(A)_i$, so that we want B to be $\iota^n{}^{\text{``}}\bigcup(A)_i$. Thus what we want is $proj(B, X)$.

So we need to know that $(A)_i$ exists. The Hailperin $(n+1)$-tuple $\langle x\vec{y}\rangle$ is the pair $\langle \iota^{2n-2}{}^{\text{`}}x, \langle \vec{y}\rangle\rangle$ so all we need is a function giving us the domain of A, a function giving us the range of A, and something to peel off the

$(2n-2)$ ιs, namely $2n-2$ applications of the axiom of sumset. The rest follows easily by induction on n. To illustrate:

$$(A)_1 = \bigcup^{n} rn(A)$$

$$(A)_2 = \bigcup^{n'} rn(dom(A))$$

$$\vdots$$

$$(A)_i = \bigcup^{n''} rn(dom^i(A))$$

where n, n', n'' depend on the length attributed to the tuples in A. This tells us we will need $dom(A)$ for the domain of A, $rn(A)$ for the range of A, and $\bigcup A$.

To deal with restricted universal quantifiers we can express them as restricted existential quantifiers flanked by negations, so it will be sufficient to deal with negations. For this we will need relative complement (set difference). There is a slight complication in that we seek the complement of $\{\langle\vec{x}\rangle : \Phi(\vec{x})\} \cap A$ not in A itself, but in the subset of A containing all the tuples of the right kind. But this subset can be assembled as a cartesian product of things like ι^j"$(A)_i$. So we will also need cartesian products. Notice that (since we are using Hailperin tuples) we need the axiom giving ι"x to form the product $X \times Y \times Z$, for this will be ι^2"$X \times (Y \times Z)$, and so on for tuples of greater length.

Now for the quantifier-free case.

We have assumed that our stratified Δ_0 formula has been formulated as a disjunction of conjunctions. We can use $A \cup B$ (pairwise union) to take unions of finitely many sets of n-tuples. So all that remains to show is that we can form cofinitely many versions of

$$\{\langle\vec{x}\rangle : \Phi(\vec{x})\}$$

when Φ is a conjunction of atomic and negatomic formulae. This reduces to the problem of forming versions of

$$\{\langle\vec{x}\rangle : \bigwedge_{i \in I} \phi_i(\vec{x})\} \cap A$$

where the ϕ_i are atomic or negatomic. This time we use $A \cap B$ to take the intersections of appropriate sets of n-tuple.

What is an appropriate set of n-tuples? Clearly we can think of ϕ_i as a conjunction of atomic or negatomic formulae, so that we are after a finite intersection of things like

$$\{\langle\vec{x}\rangle : \alpha(\vec{x})\} \cap A$$

where $\alpha(\vec{x})$ is atomic or negatomic. Now there are only two kinds of atomic wffs, so $\alpha(\vec{x})$ can only be $x_i \in x_j$ or $x_i = x_j$ or their negations. Clauses like $x_k \in A_j$ (or $A_j \in x_i$) where 'x_k' is a variable not mentioned in $\alpha(\vec{x})$ are captured by doctoring A so that it is a cartesian product whose kth factor is either A_j (or $\{u \in A_n : A_j \in u\}$ for some other A_n) or an intersection of such terms.

This tells us that we need all versions of the identity relation restricted to a set x. It is easy to check that with Wiener–Kuratowski ordered pairs this is just $\iota^2``x$, which we already have. We will also need a set of ordered pairs coding the restriction of the membership relation

$$\{\langle\iota`x, y\rangle : x \in y \wedge x \in A_1 \wedge y \in A_2\}$$

and its converse.

To be sure we can get all the versions of the fake epsilons we need we will have to have available the operation $RUSC$ from page 9 ($RUSC(R) = \{\langle\iota`x, \iota`y\rangle : \langle x, y\rangle \in R\}$).

Now we have to build up from these two sorts of sets of pairs up to the sets of n-tuples we need to take unions of. This is where it would be very nice to have the freedom to manipulate sets of n-tuples in the way alluded to at the outset. As it is, we have to be very careful. We extend the pairs to n-tuples by hanging dummy variables on either the beginning or end of the pair, or sticking them in the middle. It is important to do these in the right order. First we have to put on the end all the extra coordinates to be added there. Then we insert stuff in the middle, and finally add the ordered pairs that have to be put on the front.

How do we put extra entries on the end? Clearly it will be sufficient to be able to catenate an n-tuple onto the end of a pair. So we want to be able to form

$$\{\langle x, y, \langle\vec{z}\rangle\rangle : \langle x, y\rangle \in A_1 \wedge \langle\vec{z}\rangle \in A_2\}.$$

Let us adopt an axiom giving us the existence of

$$\{\langle x, y, z\rangle : \langle x, y\rangle \in A_1 \wedge z \in A_2\}.$$

Let us call this $cat(A_1, A_2)$. It will turn out that this will be sufficient. If $\vec{z}$ is of length n then the $(n+2)$-tuple $\langle x, y, \vec{z}\rangle$ is the pair

$$\langle\iota^{2n}`x, \langle\iota^{2n-2}`y, \langle\vec{z}\rangle\rangle\rangle$$

which is the triple
$$\langle \iota^{2n-2}{}^{\backprime}x, \iota^{2n-2}{}^{\backprime}y, \langle \vec{z} \rangle \rangle.$$
Now $\{\langle x, y, \langle \vec{z} \rangle \rangle : \langle x, y \rangle \in A_1 \wedge \langle \vec{z} \rangle \in A_2\} = cat(RUSC^{2n-2}(A_1), A_2)$ which is a set because of *cat* and *RUSC*.

So the trick is to append *simultaneously* all the extra coordinates that have to be appended.

Now we have to consider inserting coordinates in the middle. We will show that we can extend an n-tuple to an $(n+1)$-tuple by putting in a new entry immediately after the old first entry. We award ourselves another operation:

$$insert(A_1, A_2) = \{\langle x, y, z \rangle : \langle x, z \rangle \in A_1 \wedge y \in A_2\}.$$

We want to be sure that we can infer the existence of

$$\{\langle x, y, \vec{z} \rangle : \langle x, \vec{z} \rangle \in A_1 \wedge y \in A_2\}$$

from this. Now $\langle x, y, z \rangle$ is $\langle \iota^{2n}{}^{\backprime}x, \iota^{2n-2}{}^{\backprime}y, \vec{z} \rangle$ when $\vec{z}$ is of length n, so what we want is $insert(A_1, \iota^{2n-2}{}^{\backprime\backprime}A_2)$.

Adding coordinates at the beginning is easy (and this time we add them one at a time) using $\times$. For this we will need to be sure that if A is a set of n-tuples then for each $k \leq n$ the the set $(A)_k$ of kth coordinates of n-tuples in A is also a set, for each such k. But this has already been done.

Finally we are left with the atomic cases, for which the only new operations we need are $A_1 \cap B^{\backprime}A_2$, the existence of the identity function restricted to A, and the existence of singletons.

Thus we need $RUSC(R)$; $x \times y$; $x - y$; $x \cap y$; $\iota^{\backprime\backprime}x$; $x \cup y$; $\bigcup x$; $proj(x, y)$; $dom(x)$; $rn(A)$; $insert(x, y)$; $cat(x, y)$; $id \cap x \times y$; $\iota^{\backprime}x$; $\{\langle \iota^{\backprime}u, v \rangle : u \in v \in x\}$; $\{u \in y : x \in u\}$. This is certainly not the best we can do. We do not need *proj* once we have *dom*, *rn*, and $\cap$; given either *dom* or *rn* we can get the other if we have $\{\langle y, x \rangle : \langle x, y \rangle \in A\}$; we can derive $x \cap y$ as $x - (x - y)$; if we have pairing instead of existence of $\iota^{\backprime}x$ then we can drop $x \cup y$, for $x \cup y = \bigcup\{x, y\}$. Indeed if we are willing to make use of our knowledge of the internal structure of Wiener–Kuratowski ordered pairs (this is "implementation-sensitive" and not politically correct, but Hailperin [1944] does it; see below) then we can make other simplifications: $rn(A) = \bigcup(\bigcup A \cap \iota^{\backprime\backprime} \bigcup\bigcup A)$ and the restriction of the identity relation to $x \times y$ is just $\iota^2{}^{\backprime\backprime}x \cap \iota^2{}^{\backprime\backprime}y$.

Thus we can drop *dom*, *rn*, $id \cap x \times y$, *proj*, $x \cap y$, $\iota^{\backprime}x$, $x \cup y$; and adopt $\{x, y\}$ and $cnv^{\backprime}A = \{\langle y, x \rangle : \langle x, y \rangle \in A\}$.

A word is in order on the composition of stratified rudimentary functions. Suppose $F_1(x, y)$, $F_2(x, y)$, and $F_3(x, y)$ are stratified rudimentary functions of two variables, with $F_1(x, y)$ homogeneous (i.e. "$x = F_1(x, y)$"

is stratified) but F_2 and F_3 are not homogeneous. (For example, suppose "$x = F_2(x, y)$" and "$x = F_3(x, y)$" are not homogeneous but "$\{x\} = F_2(x, y)$" and "$\{x\} = F_3(x, y)$" are). What are we to make of the function $\lambda x.F_1(F_2(x, x), F_3(x, x))$?

This is a more complicated version of the question that arises with the (rudimentary) function that returns $x \cup \{y\}$ on being given x and y. What are we to make of $\lambda x.(x \cup \{x\})$? We want this to be stratified rudimentary within the meaning of the act. The point is that any structure closed under the (rudimentary) function that returns $x \cup \{y\}$ on being given x and y is also closed under $\lambda x.(x \cup \{x\})$.

Conclusions and remarks This result has various consequences. The first and most obvious is that *NF* is finitely axiomatizable (say that V is a set and is closed under the list of operations above), but we knew that already. (See page 26.) Rather more important is the possibility of defining an analogue of the constructible universe using the stratified functions presented here instead of the usual set of rudimentary functions. It is not quite clear what to do at limit stages in the construction, and this has to be thought about. The axiom of choice holds in L because there is a uniform Σ_1 well-order of the universe ("I am constructed earlier than you") which is coded in L itself since L satisfies full comprehension. Prima facie there is no reason to suppose that the version of L we construct with the finite basis of functions revealed here will be a model of unstratified comprehension. Indeed this is where the delicacy of the correct definition of the stratified–constructible hierarchy at limit stages arises. One would expect that it should be possible to define it so that the result is not a model of full Δ_0 separation. We know that the canonical well-ordering of L is—although definable—not (apparently) definable by a stratified expression. Accordingly we do not expect the stratified version of L to be a model of *AC*. Quite what this structure does satisfy is an exciting question on which work has not even begun. To be sure that this programme is not trivial we need to know that that the full Δ_0 separation scheme is not derivable from the stratified Δ_0 separation scheme. (If it were, *NF* would be inconsistent!) Friederike Körner has pointed out to me a simple proof that it is not so derivable. If it were, we would be able to show in *KF* that every set is strongly cantorian but it is a simple matter to construct an Ehrenfeucht–Mostowski model of *KF* admitting a permutation model in which there is a finite set that extends its own power set.

We can also use KF to improve a reduction of the consistency problem of *NF* due to Boffa. He remarked in [1988] that if *ZF* has a model with an $\in$-automorphism which maps x onto something the same size as $\mathcal{P}`x$ for some x, then *NF* is consistent. If we want a converse, then the apparent weakness of *NF* suggests we should find a version of this assertion with *ZF* replaced by something weaker. The appropriate weaker system is KF.

REMARK 2.3.28 *NF is consistent iff there is a model M of* KF *with an automorphism π and $a \in M$ such that $M \models \overline{\overline{\pi\text{‘}a}} = \overline{\overline{\mathcal{P}\text{‘}a}}$.*

REMARK 2.3.29 *If NF is consistent, then there is a model $M \models$* KF *with automorphism π such that $\pi\text{‘}a = \mathcal{P}\text{‘}a$ for some $a \in M$ (not just $\overline{\overline{\pi\text{‘}a}}=\overline{\overline{\mathcal{P}\text{‘}a}}$).*

As Boffa points out in his paper, no such model M of *ZF* can exist.

It is worth noting that Boffa's reduction does not depend on the a such that the automorphism moves a to something the same size as $\mathcal{P}\text{‘}a$ being *transitive.* Since, in *ZF*, "every set is the same size as a transitive set" implies that every infinite well-founded set has a countable partition, it is a non-negligible (but weak) version of *AC*, and adding the condition that a is transitive is not obviously cheap. Kaye has remarked that, if a is transitive, then the resulting model of *NF* admits a well-founded extensional relation. The status in *NF* of this last assertion is problematical and has already been mentioned (theorem 2.1.13). This matter deserves further investigation.

There is an analogous reduction of the consistency problem of KF+$\exists$*NO* to the problem of finding a model of KF with an automorphism σ and a set x so that $\sigma^n\text{‘}(\aleph\text{‘}\ \overline{\overline{\mathcal{P}^n\text{‘}x}}\ )$ is constant for all n.

2.4 Subsystems, term models, and prefix classes

Hinnion in [1972] showed that every finite binary structure can be embedded in every model of *NF*. One can regard Hinnion's theorem rather as a fact about provability of members of selected prefix classes: it says that every satisfiable $\exists_1$ sentence is provable in *NF*. Several theorems of this kind concerning *NFO*, NF_2, and various prefix classes have since been proved and several more conjectured. Some of them involve term models. The existence of term models for a theory is obviously intimately connected to the question of whether or not it has the existence property.

DEFINITION 2.4.1 *T has the existence property iff whenever $T \vdash (\exists x)(\Psi(x))$ then $T \vdash \Psi(t)$ for some term t of T. By a "term" in this context we will mean a closed set abstract.*

I am indebted to Martin Hyland for drawing my attention to the question of whether or not *NF* has the existence property. The answer to Hyland's question is short and unsatisfactory.

REMARK 2.4.2 *NF does not have the existence property.*

Proof: Consider $\phi\ :\ (\exists x)(\forall y)(y = \iota\text{‘}y \to x = \iota\text{‘}x)$. This is logically valid and therefore a theorem of *NF*. But no term t can *provably* be a witness to the existential quantifier in ϕ for then, in any model with Quine atoms (and we can always add Quine atoms by means of a permutation as we shall see in proposition 3.1.6), we would have $t = \iota`t$, where t was the term that was a witness to ϕ. But it is easy to show that no term can denote a Quine atom. This ϕ is a counterexample to the existence property. ■

The proof of this result reveals that I have taken Hyland too literally and that it is an answer to the wrong question. What is unsatisfactory about this proof is first that it makes essential use of the full power of classical logic—$(\exists x)(\forall y)(y = \iota`y \to x = \iota`x)$ is *not* intuitionistically correct—and second it uses formulae that are not *stratified*. Indeed all known counterexamples have *both* of these undesirable features. Clearly the *right* questions are as follows.

1. If we have a proof in *intuitionistic NF* of $(\exists x)(\Psi(x))$ for some *stratified* Ψ, can we produce a witness?
2. Does intuitionistic *NF* admit a normalization theorem?

These questions are obviously at least as hard as the consistency question for *NF*, so (as for that question) negative answers may be easier to find than positive! The natural candidate for a non-normalizable proof in *NF* is the proof of the axiom of infinity, but this contains extensive use of the axiom of extensionality and is far too large to examine by hand.

In general there is very little known about the proof theory of *NF*. Readers who wish to pursue this should read the works of Crabbé itemized in the bibliography, and perhaps pick through the works of Kuzichev for promising morsels.

We shall start on the positive results by proving two theorems that extend Hinnion's result, and close with a few conjectures.

THEOREM 2.4.3 *Every countable binary structure can be embedded in the term model for NFO.*

Proof: The term model of *NFO*can be thought of as the algebra of all words in *NFO*operations reduced by *NFO*-provable equations. A proof that this quotient is well-defined can be found on p. 376 of Forster [1987b].

Let us suppose we are trying to construct an embedding i from a structure $\langle \mathbb{N}, R \rangle$ into the term model so that $n\ R\ m \longleftrightarrow i`n \in i`m$. We will need an infinite supply of distinct $x \in x$ and distinct $y \notin y$, and such a supply can easily be found with the help of the B function. Let the nth *left* object be $B^n`V$ and the $n$th *right* object be $B^n`\Lambda$. By remark 2.1.11, all left objects are self-membered and no right objects are. We will construct i by recursion on $\mathbb{N}$. First $i`1$ is the first left or right object according to $1\ R\ 1$. At later stages $n$ we have to construct $i`n$ as an *NFO* term. Let O_n be the $2n$th left object, if $n\ R\ n$, or right object if not. Then $i`n$ will be obtained from O_n by adding and removing only finitely many things.

We have four sets to consider:
A: $\{i`m : m < n \wedge m\, R\, n\}$
B: $\{i`m : m < n \wedge \neg(m\, R\, n)\}$
C: $\{i`m : m < n \wedge n\, R\, m\}$
D: $\{i`m : m < n \wedge \neg(n\, R\, m)\}$.

$i`n$ must extend $A$, be disjoint from $B$, belong to everything in $C$, and to nothing in $D$. So our first approximation is $(O_n - B) \cup A$. For each $i`k \in C$ we want $i`n \in i`k$. Now $i`n \in O_k \longleftrightarrow B^{-1}`O_k \in i`n$, so we can determine the truth value of '$i`n \in O_k$' (at least) by putting $\iota`(B^{-1}`O_k)$ into $i`n$ or not. It will follow from

$$(\forall n \in \mathbb{N})(\forall k < n)(i`n \in i`k \longleftrightarrow i`n \in O_k)$$

that this actually determines the truth value of '$i`n \in i`k$' as well. Consider a notion of rank of *NFO* terms as the depth of nested occurrences of 'B'. To get $i`k$ from $O_k$ we remove and add only odd rank items $\neq i`i$ or any $i`j$ with $j < k$: neither can affect $i`n$. ■

THEOREM 2.4.4 *Every $\forall_2$ sentence is either true in all sufficiently large models of* TST *in which the bottom type is finite, or false in all sufficiently large models of* TST *in which the bottom type is finite.*

Let us call these *finitely generated* models. First we show that every countable model of TST is a direct limit of finitely generated models of TST. Despite appearances, this is not at all obvious: the obvious embedding produces a direct limit where each type is not a boolean algebra but a copy of V_ω. To prove this we will need a sublemma.

SUBLEMMA 2.4.5 *Let $\mathbf{B}$ be a countable atomic boolean algebra which is a direct limit of $\langle \mathbf{B}_n : n < \omega \rangle$, a sequence of finite subalgebras of $\mathbf{B}$. Let $\mathbf{B}'$ be a countable atomic subalgebra of $\mathcal{P}`\mathbf{B}$ containing all singletons. Then there is a sequence $\langle \pi_n : n < \omega \rangle$ of partitions of $\mathbf{B}$ such that, for each i,*

π_i refines π_{i-1}
$\pi_i \subseteq \mathbf{B}'$
$\mathbf{B}_i$ is a selection set for π_i.

Further, if we let ${\mathbf{B}'}_i$ be the subalgebra of $\mathbf{B}'$ generated by π_i (so that $\mathcal{P}`\mathbf{B}_i \simeq {\mathbf{B}'}_i$) then the union of the ${\mathbf{B}'}_i$ is $\mathbf{B}'$.

$\mathbf{B}'$ is then the direct limit of the ${\mathbf{B}'}_i$ where the injections are inclusion.

Proof: The recursive construction we use involves iterating a step whose execution may consume several indices, so we must distinguish between *initial* and *transition* indices; 1 is an initial index. We have a well-ordering $\langle x_n : n < \omega \rangle$ of $\mathbf{B}'$.

Let i be an initial index. We have a partition π_{i-1} (π_0 is the $\mathbf{1}$ of $\mathbf{B}$); we want to refine it to π_i for which $\mathbf{B}_i$ will be a selection set; and we want the corresponding boolean algebra $\mathbf{B}'_i$ to contain x_j, where x_j is the first x not already a member of a $\mathbf{B}'_k$. So we want x_j to be a union of elements of π_i. That way, x_j is an element of $\mathbf{B}'_i$ and will appear in the direct limit, which will therefore be $\mathbf{B}'$ as desired. If we can do this, then we have ensured that every x_j appears in some $\mathbf{B}'_i$ or other. Let us say b *crosses* x iff $b \cap x$ and $b - x$ are both non-empty. We have to refine π_{i-1} to a partition whose every element is either included in, or disjoint from, x_j, that is to say, to a partition with no elements that cross x_j. If we can do this, we choose π_i to be such a partition. The difficulty is, every element of the refined partition must contain precisely one element of $\mathbf{B}_i$. A problem will arise if there is an element b of π_{i-1} that crosses x_j, but of the new points of $\mathbf{B}_i$ that we can use, all (and there may not be any in b anyway) belong to the same part, be it $b - x$ or $x \cap b$, that contains the representative of $\mathbf{B}_{i-1}$. If there is such a b we must continue considering $\mathbf{B}_{i+1}$, $\mathbf{B}_{i+2}$, etc., until we find a member of $\mathbf{B}_{i+n}$ which does belong in the half of b not containing the representative of $\mathbf{B}_{i-1}$. Then we can partition b into bits each of which is included in or disjoint from x, and each of which contains precisely one point of $\mathbf{B}_{i+n}$. This will happen for some finite n because, by hypothesis, everything in $\mathbf{B}$ sooner or later appears in one of the $\langle \mathbf{B}_n : n < \omega \rangle$. We can easily fill in the details of $\pi_i, \ldots, \pi_{i+n-1}$ so that the appropriate subalgebras are selection sets for them (these are the transition indices). When we have finished this, the next index is an initial index. ■

LEMMA 2.4.6 *Every countable model of* TST *is a direct limit of all finitely generated models of* TST.

Proof: Let M, a countable model of simple type theory, be a family $\langle \mathbf{B}_n : n < \omega \rangle$ of countable atomic boolean algebras, where $\mathbf{B}_{n+1}$ is a countable atomic subalgebra of $\mathcal{P}‘\mathbf{B}_n$. Let $\mathbf{B}_1$ be a union of an ω-sequence $\langle \mathbf{B}_1^i : i < \omega \rangle$. We then invoke sublemma 2.4.5 repeatedly to obtain, for each n, families $\langle \mathbf{B}_n^i : i < \omega \rangle$ of subalgebras and $\langle \pi_n^i : i < \omega \rangle$ of partitions as above. Now consider the structures $\langle \langle \mathbf{B}_n^i : n < \omega \rangle, \in \rangle$ for $i < \omega$. We have constructed the $\mathbf{B}_n^i$ so that $\mathbf{B}_n^{i+1}$ is an atomic boolean algebra whose atoms are elements of a partition for which $\mathbf{B}_n^i$ is a selection set. Thus, if we want to turn the $\langle \mathbf{B}_n^i : n < \omega \rangle$ into a model of simple type theory the obvious membership relation to take is $\in$ itself. They are models of simple type theory without the axiom of infinity, and by construction their direct limit is pointwise the nth type of M, so the direct limit is M as desired. ■

We now return to the proof of the main theorem. Suppose Φ is $\exists_2$ and has arbitrarily large finitely generated models. Then it has a countable model. This model is a direct limit of all finitely generated models, and accordingly would satisfy $\neg\Phi$ (which is $\forall_2$) if $\neg\Phi$ had arbitrarily large finitely generated models. This is impossible so, if Φ is $\forall_2$, then Φ and $\neg\Phi$ cannot both have arbitrarily large finitely generated models. ■

This is the closest I have got to proving that *NFO* decides all stratified $\forall_2$ sentences, which I conjecture to be true. If it is, it would be the best possible. We cannot extend it to unstratified $\forall_2$ sentences: as we shall see in section 3.1.1 '$(\forall x)(\exists y)(y \in x \longleftrightarrow y \neq x)$' is consistent and independent of *NF*. It is an open problem to find the simplest quantifier prefix class that contains a *stratified* sentence undecidable in *NF*, and likely to remain so until we discover a technique for showing the independence of stratified expressions that works as well as permutations (chapter 3) work for unstratified ones. A good bet would be $str(\exists_3)$, for the assertion that V can be ordered is $str(\exists_3)$, and there seems to be no obvious reason why it should be inconsistent, while its independence would follow from the existence of a term model for *NF*, as we have seen in section 2.1.3.

Lemma 2.4.7 *Any $\forall_2$ sentence true in some model of NFO is true in the term model of NFO.*

Proof: First we minute the following facts about the term model for *NFO*.

(1) It can be embedded in every model of *NFO*. (2) It is a model of the axiom of extensionality. (3) Its diagram is a complete and decidable theory.

(1) is obvious. To show (2) and (3) it will suffice to show that every *NFO* word that does not provably transform to Λ has an *NFO* word as a member. To do this we need a notion of *rank* of *NFO* terms, which will be the depth of nesting of B and ι. We show easily that any question $t_i = t_j$? or $t_i \in t_j$? can be reduced to a boolean combination of similar questions about terms of lower rank. (See Forster [1987b], proposition 3, p. 376, for details.)

Now suppose '$(\forall \vec{x})(\exists \vec{y})(\Phi(\vec{x}, \vec{y}))$' is true in some model M of *NFO*. Let us work in this model, bearing in mind that there is no serious loss of generality because (1) and (3) above imply that we get the same structure of terms no matter what model we find them in. So it is certainly true that, for all (*NFO*-)terms $\vec{t}$, we have $(\exists \vec{y})(\Phi(\vec{t}, \vec{y}))$, and the task is to show that witnesses for the '$(\exists \vec{y})$' can be found among the terms. We know that atomic sentences in Φ need never be of the form '$y_j \in t_i$', because any such atomic wff can be expanded until it becomes a boolean combination of atomic wffs like '$y_i = t_j$', '$y_j \in y_i$', and '$t_j \in y_i$'. Then we can recast the matrix into disjunctive normal form. We know that '$(\forall \vec{x})(\exists \vec{y})(\Phi(\vec{x}, \vec{y}))$' is true so there is at least one disjunct that does not trivially violate the theory of identity. This disjunct is a conjunction of things like '$y_i = t_j$', '$y_j \in y_i$', and '$t_j \in y_i$' and their negations, atomic wffs not containing any

$\vec{y}$ having vanished since they are decidable. We now have to find ways of substituting *NFO* terms $\vec{w}$ for the $\vec{y}$ to make every conjunct in the disjunct true. First we reactivate the idea of the sequence of *left objects* (which are all self-membered) and *right objects* (none of which are) from the proof of theorem 2.4.3 and the notion of rank of *NFO* terms in terms of the depth of nesting of B. We construct witnesses for the $\vec{y}$ in the way we constructed values of the function i in the proof of theorem 2.4.3. Let n_0 be some fixed integer greater than the rank of all the t_i that appear in our disjunct. We know of y_0 that it is to have certain ts as members and certain others not. We construct a word w_0 which is the n_0th left member (if '$y_0 \in y_0$' is a conjunct) or the n_0th right object (otherwise) $\cup$ (the tuple of t_i such that '$t_i \in y_0$' is a conjunct) minus (the tuple of t_j such that '$t_j \notin y_0$' is a conjunct). From here on, we construct words w_i to be witnesses for y_i in exactly the same way as we proved theorem 2.4.3. ■

Does *NF* have a term model? We have seen that *NFO* has a term model, and in Forster [1987b] I show that *NF*∀ has one too. In any term model, every set is symmetric. Can *NF* at least have a model in which every set is symmetric? Let M be such a model and consider $(J_0)^M$, the full symmetric (beware pun) group of all internal permutations of M. For $\sigma \in (J_0)^M$, σ is n-symmetric iff it is in the centralizer $C_{(J_0)^M}((J_n)^M)$, and so, since every permutation in M is indeed symmetric, we have $(J_0)^M = \bigcup_{n<\omega} C_{(J_0)^M}((J_n)^M)$. Now Macpherson and Neumann [1990] proved that no symmetric group can be a union of an ω-sequence of centralizers. Unfortunately their proof uses a diagonal construction that depends on the ω-sequence of centralizers being a *set* of the model, so is not applicable here. The question of whether or not *NF* has a term model remains open.[39]

In term models and models in which every set is symmetric we can show that the membership relation restricted to small sets is well-founded—if we define "small" properly!

PROPOSITION 2.4.8 *The membership relation restricted to symmetric finite sets is well-founded.*

Proof: If not, there is a sequence $x_1, x_2, x_3, \ldots$, with $x_{n+1} \in x_n$ for all n, and satisfying the following. x_1 is n-symmetric and finite; x_2 has only finitely many $(n-1)$-copies (they are all in x_1); x_3 has finitely many $(n-2)$-copies (because they are all packed into the finitely many $(n-1)$-copies of x_2); x_4 has only finitely many $(n-3)$-copies (because they are all packed

[39]Nothing is gained by recreating Macpherson and Neumann's proof in *NF* + the existence of the set $\{C_{(J_0)^M}((J_n)^M) : n \in \mathbb{N}\}$ to show that it has no term model, because we can prove directly that $\{C_{(J_0)^M}((J_n)^M) : n \in \mathbb{N}\}$ cannot be the denotation of a closed term anyway.

into the finitely many $(n-1)$-copies of x_2); and so on down to some object that has only finitely many 1-copies, which is impossible. ■

I proved proposition 2.4.8 after Friederike Körner's [1994] construction of a model of *NF* in which the membership relation restricted to finite sets is well-founded. It is sensible to ask if this can be proved for larger sets too. Let us say I is a *notion of smallness* if

1. Any subset of an I thing is also I.
2. Any union of I-many I-sets is I.
3. V is not I.

Then the proof of proposition 2.4.8 generalizes to a proof that $\in$ restricted to small symmetric sets is well-founded.

NFO, NF_3, and *NFU* are not the only subsystems of *NF* that have been studied. Others are discussed in Boffa and Casalegno [1985], Forster [1987b], Crabbé [1982a], and in Oswald (all *op. cit.*).

2.5 The converse consistency problem

The converse consistency problem is the problem of obtaining models or consistency proofs for set theories in the *ZF* tradition, working within a theory in the *NF* tradition. Two sorts of question fall under this heading. On the one hand, one can work within a theory in the *NF* tradition whose consistency is unknown, and try to prove (as it might be) $Con(ZF)$ in it. On the other, one can consider theories in the *NF* tradition for which methods of constructing models are known (which really means *NFU* and NF_3) and see if the method can be jigged to build a model big enough to contain a model of (for example) *ZF* inside it. I find the first question much more interesting than the second, for the second asks how much well-founded structure we can stuff into a model with a universal set, whereas the first asks how much of this structure arises spontaneously from the comprehension principles themselves. Only negative results for the second question would be interesting, and there are none known (or expected). Positive results for the second kind of question are to be found in chapter 4.

All work done on attempts to reconstruct *ZF* and its kin in *NF* is in Hinnion [1975, 1976, 1979, 1990], Orey [1955, 1956], Rosser [1956], and Forster [1976]. This section is largely a résumé of these essays.

In [1955] and [1956], Orey is interested in the question of what halfway-sensible axioms one has to add to *ML* to get a natural proof of the consistency of *ZF*. What he does is add enough induction and recursion over ordinals to enable the Gödel construction of L to go through. The general feeling nowadays seems to be that it is simplest to stick to *NF*, and accordingly we will not discuss Orey's construction here. Interested readers may try Orey [1955, 1956] which are self-contained. Hinnion and Forster work in

NF, and are (roughly) interested in seeing what one can do in *NF* without adding any axioms at all if this can be avoided. This is more helpful if we are interested in determining the consistency strength of *NF*. In fact both techniques used here amount to no more than showing how to interpret Z into *NF*, and the set existence axioms available in *NF* will determine whether we get a relative or absolute consistency result.

2.5.0.1 *Hinnion's study of well-founded extensional relational types* Let *BF* ("*B*ien *F*ondée") be the set of relational types of well-founded extensional relations that have a unique maximal element. These relations clearly look like the $\in$-graphs of transitive sets. There is an obvious membership relation between them (which we shall write "E"), and there is a type-raising operation on these relational types (beware ambiguity of the word "type"!) which we will write T for conformity with the relation on NC and NO. Evidently T is an isomorphism from $\langle BF, E\rangle$ to some proper subset of itself, but we shall say little about it since it does not figure largely in Hinnion's construction. Since E is well-founded, we can prove things by induction on it, and, in an exact analogue of remark 2.2.2, we have

REMARK 2.5.1 *The relational type of E restricted to $\{\mathcal{A} : \mathcal{A}\ E\ \mathcal{B}\}$ is $T^2\mathcal{B}$.*

We prove this by a straightforward induction. ■

It is also fairly clear that E-analogues of the sequence of V_α well known from *ZF* will appear. Quite how many is a good question. Now since 'E' is homogeneous, any set-theoretical expression *whatever* with '$\in$' replaced by 'E' is *stratified*; so, if we can show that the E-analogue of $V_{\omega+\omega}$ exists, then we will have no problem showing that it is an E-model for Z. Now, since the E-analogue of $V_{\omega+\omega}$ exists iff the E-analogue of $V_{\omega+n}$ exists for each $n \in \mathbb{N}$, and $V_{\omega+n}$ exists iff the cardinal $\beth_n$ exists, we are back to the familiar problem seen in section 2.2. As we saw there, AxCount$_\leq$ certainly proves the existence of all alephs and beths with finite subscripts, so it will do the trick for us here. Thus we can prove the following.

PROPOSITION 2.5.2 $NF + AxCount_\leq \ \vdash \ Con(Z)$.

A modification enables Hinnion [1979] to weaken AxCount$_\leq$ in the antecedent to "$\aleph_\omega$ exists", but it is easier to prove this by the technique of the following section. ■

Having drawn relational types of well-founded extensional relations to our attention, Hinnion then does not do a great deal with them. There is (presumably) an analogue of Specker's refutation of *AC* and Henson's lemma 2.2.10, though nobody seems to have bothered to prove (or even state) it. In [1975] Hinnion also proved that it is consistent with respect to *NF* to suppose that Mostowski's collapse lemma holds for well-founded extensional relations whose domains are strongly cantorian. We will see this in proposition 3.1.8.

2.5.0.2 *Models in the ordinals* Constructible sets are defined by a recursive construction over the ordinals. There is nothing in principle to prevent us constructing instead a sort of membership relation E which makes $\langle On, E\rangle$ look like $\langle L, \in\rangle$. Indeed there is nothing to prevent one performing this construction on any well-ordering whatever. Such constructions have recently come to be called "*L-strings*". In Forster [1976] I outlined this construction of an L-string (though they were not then called that) over NO and claimed that by this means one could prove that Z was consistent relative to NF. Rosser [1956] seems to have had the same idea. The execution of this programme is much more complicated than I at first thought. Of course if there is a well-ordering of length ω_ω then we will certainly get a model of Z and thereby a proof of $Con(Z)$ inside NF, but the aim of the project was to show that the model corresponding to the cantorian ordinals satisfied Z, and this has still not been demonstrated to everyone's satisfaction: problems with extensionality look insuperable. Of course in $NF+$ AxCount$_\leq$ we can prove the existence of ω_ω and thence the consistency of Z, but this is merely another proof of proposition 2.5.2.

3

PERMUTATION MODELS

The first thing to say about this permutation method is that it is *not* the permutation method of Fraenkel and Mostowski. It is due to Rieger [1957] and Bernays [1954] and was first applied to *NF* by Scott [1962]. It is familiar to students of *ZF* as the technique used in the standard proof of the independence of the axiom of foundation. If we have a model $\langle V, \in \rangle$ of (say) *ZF*, we can take the transposition $\sigma = (a, \iota\text{'}a)$ for some fixed arbitrary a. Then we define $x \in_\sigma y$ by $x \in \sigma\text{'}y$, obtaining a new model $V^\sigma = \langle V, \in_\sigma \rangle$. In V^σ if $y \in a$ then $y \in \iota\text{'}a$, so $y = a$. That is to say, a is identical to its own singleton and, in general, if σ swaps x and $\iota\text{'}x$, then x is a Quine atom in V^σ.

The reason for an extended treatment of this technique here is the great importance that it assumes in set theories with a universal set, for many of which it is the only independence technique available. In general, if $\langle V, R \rangle$ is a model for the language of set theory, and π is any (possibly external) permutation of V, then we say $x \; R_\pi \; y$ iff $x \; R \; \pi\text{'}y$. $\langle V, R_\pi \rangle$ is a *permutation model* of $\langle V, R \rangle$. We call it V^π and we speak of V as the *base set* for all these permutation models. Another way of describing this move is to define Φ^π as the result of replacing every atomic wff $x \in y$ in Φ by $x \in \pi\text{'}y$. ($=$ is not affected in this move to Φ^π.) We then discover that the result of our definitions is that $\langle V, R \rangle \models \Phi^\pi$ iff $\langle V, R_\pi \rangle \models \Phi$. The status of the variable place occupied by 'π' is not totally determined by this, and some discussion is in order. At various later stages, we will take π to be an internal permutation (recall the notation 'J_0' for the set of all (internal) permutations of V in this connection) or an (external) automorphism of $\langle V, \in \rangle$.

In the first case "$(\forall \pi)\Phi^\pi$" is to be understood as "$(\forall x \in J_0)\Phi^x$" (prepare to think of this as "$\Box\Phi$") and the existential case similarly. In section 3.1.1.1 we discuss a modal treatment of these two notions, show that (at least if we are working in *NF*) the interpretation is sound, and characterize the logic thus interpreted. Our motive for wishing to treat permutation models modally is of course that we might wish to consider *iterating* the act of proceeding to a permutation model. This is more complicated than one might think. '$\exists \pi \in J_0(\phi^\pi)$' is not a formula bearing any resemblance to ϕ, for '$(x \in_\pi y)^\tau$', when written out in primitive notation, is an extremely long formula (remember there is talk of functional application in there, and we are using Quine ordered pairs). So making sense of '$(x \in_\pi y)^\tau$' involves more than just composing two permutations: it involves presenting π as an argument to τ, and this of course can happen only if π is a *set*. We will certainly be restricting our attention throughout this chapter to permutations that are setlike (otherwise we have no control over M^τ), but we will at different times make different assumptions about additional properties τ may have, such as being a set of the model in question (if we make this assumption, we can iterate and obtain a modal theory) or being an $\in$-automorphism. These assumptions are important and the reader is urged to note which one is in effect at any time.

A natural move is to consider the family of all permutation models obtainable from one model. A set theorist would then be inclined to ask questions about the structure of embedding relations between these permutation models: this material will be tackled in section 3.1.1.4.

The rest of this section is devoted to a proof of its principal result, theorem 3.0.4, a preservation theorem about permutation models due to Henson, Pétry, and the author, but we shall need to remind ourselves about a few definitions before we can state it, and some more before we can prove it.

Let $\mathcal{M}$ be a structure for the language of set theory which is also a model of the axiom of extensionality. Let τ be a permutation of $\mathcal{M}$. Since each $x \in \mathcal{M}$ can be regarded as an (external) subset of $\mathcal{M}$, it follows that τ gives rise to a map $x \mapsto \tau``x$: $\mathcal{M} \to \mathcal{P}`\mathcal{M}$. Recall the definition of $j$ as $\lambda f \lambda x.(f``x)$ so that this map is notated '$j`\tau$'. Then $j`\tau$ may or may not be onto $\mathcal{M}$, for there is in general no guarantee that the image of any subset of $\mathcal{M}$ by translation under τ is a set in the sense of $\mathcal{M}$, which is what we would need for $j`\tau$ to be a permutation of $\mathcal{M}$. (This is obviously something to do with the axiom scheme of replacement, which says that the image of a set in a function is a set. The only assumption we are making about $\mathcal{M}$ is that it is a model of extensionality.) Recall also the definition of *setlike*: τ is a setlike permutation of $\mathcal{M}$ iff, for all $n \in \mathbb{N}$, $j^n`\tau$ is a map $\mathcal{M} \to \mathcal{M}$. We note without proof that the setlike permutations form a subgroup of the symmetric group on $\mathcal{M}$. Let $\langle \mathcal{M}, \in_\mathcal{M} \rangle$ and $\langle \mathcal{N}, \in_\mathcal{N} \rangle$ be two structures for the language of set theory and models of extensionality. Let $\langle \pi_i : i \in \mathbb{N} \rangle$ be a family of maps from $\mathcal{M} \times \{i\}$ to $\mathcal{N} \times \{i\}$. $\mathcal{M} \times \mathbb{N}$ ("$\mathcal{M}^\infty$" for short) has an obvious many-sorted structure where, for x in $\mathcal{M} \times \{i\}$ and y in $\mathcal{M} \times \{i+1\}$, we say $x \in y$ iff the first component of x $\in_\mathcal{M}$ the first component of y. Hereafter we shall speak of '$\mathcal{M}_i$' instead of '$\mathcal{M} \times \{i\}$'.

DEFINITION 3.0.3 *$\langle \pi_i : i \in \mathbb{N} \rangle$ is a stratimorphism of $\langle \mathcal{M}, \in_\mathcal{M} \rangle$ onto $\langle \mathcal{N}, \in_\mathcal{N} \rangle$ iff each π_i is a bijection $\mathcal{M}_i \to \mathcal{N}_i$ and (considering π_i as the corresponding map from $\mathcal{M}$ to $\mathcal{N}$—we shall make this identification often) we have $(\forall xy)(\forall i)[(x \in \mathcal{M}_i \wedge y \in \mathcal{M}_{i+1} \wedge x \in y) \longleftrightarrow \pi_i`x \in \pi_{i+1}`y]$, where* *the unsubscripted $\in$ is the "obvious" many-sorted relation alluded to above.*

Thus, if $\langle \pi_i : i \in \mathbb{N} \rangle$ is a *stratimorphism* of $\langle \mathcal{M}, \in_{\mathcal{M}} \rangle$ onto $\langle \mathcal{N}, \in_{\mathcal{N}} \rangle$, then π_{n+1} is simply the restriction of $j`\pi_n$ to $\mathcal{M}$. If there is a stratimorphism from one structure to another we shall say that the two are *stratimorphic.*

THEOREM 3.0.4 Pétry Henson Forster. *A sentence is equivalent to a stratified sentence iff it is preserved under all permutation models where the permutation is setlike.*

The proof proceeds by a series of lemmas.

LEMMA 3.0.5 *If $\langle \mathcal{M}, \in_{\mathcal{M}} \rangle$ and $\langle \mathcal{N}, \in_{\mathcal{N}} \rangle$ are stratimorphic, then they are elementarily equivalent with respect to stratified sentences.*

Proof: $\langle \mathcal{M}, \in_{\mathcal{M}} \rangle$ and $\langle \mathcal{N}, \in_{\mathcal{N}} \rangle$ being stratimorphic means that the two many-sorted structures $\mathcal{M}^\infty$ and $\mathcal{N}^\infty$ defined above are isomorphic as many-sorted structures: that is precisely the content of the definition of stratimorphism. There is a canonical method of interpreting many-sorted theories in one-sorted theories, which the reader is assumed to know. I will call it the *canonical interpretation.* $\mathcal{M}^\infty$ and $\mathcal{N}^\infty$ are at least elementarily equivalent with respect to the many-sorted language. But every stratified sentence arises as the value of a sentence in the many-sorted language of typed set theory under the canonical interpretation; also it is not hard to see that if ϕ is a sentence of the many-sorted language then $\mathcal{M}^\infty$ satisfies ϕ iff $\langle \mathcal{M}, \in_{\mathcal{M}} \rangle$ satisfies ϕ's canonical translation. Therefore $\langle \mathcal{M}, \in_{\mathcal{M}} \rangle$ and $\langle \mathcal{N}, \in_{\mathcal{N}} \rangle$ satisfy the same stratified sentences. ■

The converse is also true, as Pétry [1982] has shown.

LEMMA 3.0.6 *If $\langle \mathcal{M}, \in_{\mathcal{M}} \rangle$ and $\langle \mathcal{N}, \in_{\mathcal{N}} \rangle$ are elementarily equivalent with respect to stratified sentences, they have stratimorphic ultrapowers.*

The proof is a relatively straightforward modification of the Keisler ultrapower lemma. Next we show the following.

LEMMA 3.0.7 *Let $\langle \mathcal{M}, \in_{\mathcal{M}} \rangle$ and $\langle \mathcal{N}, \in_{\mathcal{N}} \rangle$ be two structures for the language of set theory (both models of extensionality). Then $\langle \mathcal{M}, \in_{\mathcal{M}} \rangle$ and $\langle \mathcal{N}, \in_{\mathcal{N}} \rangle$ are stratimorphic iff $\langle \mathcal{N}, \in_{\mathcal{N}} \rangle$ is isomorphic to $\mathcal{M}^\tau$ for some setlike τ.*

Proof:

Left to right

$\langle \mathcal{M}, \in_{\mathcal{M}} \rangle$ and $\langle \mathcal{N}, \in_{\mathcal{N}} \rangle$ are stratimorphic, so let f_i be the map $\mathcal{M} \to \mathcal{N}$ induced by the map sending $\mathcal{M}_i$ 1-1 onto $\mathcal{N}_i$. Therefore, for each n, $(f_{n+1})^{-1} \circ f_n$ is a permutation of $\langle \mathcal{M}, \in_{\mathcal{M}} \rangle$. Let us call it π_n. Evidently

$\pi_{n+1} = j`\pi_n$, so π_1 is setlike. M^{π_n} is isomorphic to $\langle \mathcal{N}, \in_{\mathcal{N}} \rangle$ for any n; for instance,

$$M^{\pi_1} \models x \in y$$

iff

$$M \models x \in \pi_1`y$$

iff

$$\langle \mathcal{M}, \in_{\mathcal{M}} \rangle \models x \in (f_1)^{-1} \circ f_0`y$$

iff

$$N \models f_0`x \in f_1 \circ (f_1)^{-1} \circ f_0`y = f_0`y.$$

So f_0 is an isomorphism between M^{π_1} and $\langle \mathcal{N}, \in_{\mathcal{N}} \rangle$.

Right to left

Let τ be a setlike permutation of $\mathcal{M}$. We will construct $\langle \pi_i : i \in \mathbb{N} \rangle$ by recursion on i, where each $\pi_i : \mathcal{M} \to \mathcal{N}$. We turn them into components of a stratimorphism in the obvious way. π_0 is some arbitrary setlike permutation of $\mathcal{M}$, for simplicity's sake, the identity. Thereafter we will want to know that

$$x_n \in y_{n+1} \longleftrightarrow \pi_n`x_n \in_\tau \pi_{n+1}`y_{n+1}$$

which is to say

$$x_n \in y_{n+1} \longleftrightarrow \pi_n`x_n \in \tau\pi_{n+1}`y_{n+1}.$$

But we have

$$x_n \in y_{n+1} \longleftrightarrow \pi_n`x_n \in (j`\pi_n)y_{n+1}$$

since $u \in v \longleftrightarrow \sigma`u \in (j`\sigma)`v$ in general. So $\tau\pi_{n+1}`y_{n+1} = (j`\pi_n)y_{n+1}$, which is to say

$$\pi_{n+1}`y_{n+1} = (\tau^{-1})(j`\pi_n)y_{n+1},$$

that is

$$\pi_{n+1} = (\tau^{-1})(j`\pi_n).$$

This recursive definition of π_n shows why we have to assume τ is setlike, for only then will the recursion actually output permutations of $\langle \mathcal{M}, \in_{\mathcal{M}} \rangle$ as desired. ■

We now know that $\mathcal{M}$ and $\mathcal{N}$ are elementarily equivalent with respect to stratified sentences iff they have stratimorphic ultrapowers and that they are stratimorphic iff they are permutation models of each other. To complete the proof, we can use standard techniques for proving preservation theorems, for example lemma 3.2.1 on page 124 of Chang and Keisler [1973]. ■

It is often—indeed usually—said of the stratification discipline for avoiding the paradoxes that it is an *ad hoc* syntactic trick with no semantic underpinnings. Theorem 3.0.4 shows that this is not true. Quite what the significance is of the semantics in that preservation theorem remains to be ascertained. It is worth noting that a similar result can be proved that characterizes (among other things) the well-typed formulae of the λ-calculus (see Forster [1993]).

3.1 Permutations in *NF*

We shall start with a lemma and a definition, both due to Henson [1973a]. The definition arises from the need to tidy up Φ^τ. A given occurrence of a variable 'x' which occurs in 'Φ^τ' may be prefixed by 'τ' or not, depending on whether or not that particular occurrence of 'x' is after an '$\in$'. This is messy. Henson's insight was as follows. Suppose we have a stratification for Φ and permutations τ_n (for all n used in the stratification) related somehow to τ, so that, for each n,

$$x \in \tau`y \text{ iff } \tau_n`x \in \tau_{n+1}`y.$$

Then by replacing ' $x \in \tau`y$ ' by ' $\tau_n`x \in \tau_{n+1}`y$ ' whenever '$x$' has been assigned the subscript $n$, every occurrence of '$x$' in ' $\Phi^\tau$ ' will have the same prefix. Next we will want to know that $\tau_n$ is a permutation, so that in any wff in which '$x$' occurs bound—$(\forall x)(\ldots \tau_n`x \ldots)$—it can be relettered $(\forall x)(\ldots x \ldots)$ so that 'τ' has been eliminated from the bound variables. It is not hard to check that the definition we need to make this work is as follows.

DEFINITION 3.1.1 $\tau_0 =$ *identity*, $\tau_{n+1} = (j^n`\tau)\tau_n$.

This definition is satisfactory as long as $j^n`\tau$ is always a permutation of V whenever τ is, for each n. For this we need τ to be setlike. The trick of relettering variables that this facilitates is of crucial importance and will be used often. It is worth noting that the subscripts n are metalanguage variables and may not be bound in first-order formulae.

This gives us immediately a proof of the following result.

LEMMA 3.1.2 Henson [1973b]. *Let Φ be stratified with free variables 'x_1', ..., 'x_n', where 'x_i' has been assigned an integer k_i in some stratification. Let τ be a setlike permutation and V any model of NF. Then*

$$(\forall \vec{x})V \models (\Phi(\vec{x})^\tau \longleftrightarrow \Phi(\tau_{k_1}`x_1 \ldots \tau_{k_n}`x_n)).$$

In the case where Φ is closed and stratified, we infer that, if V is a model of *NF* and τ a setlike permutation, then

$$V \models \Phi \longleftrightarrow \Phi^\tau.$$

This was proved by Scott in [1962], but we can recognize it as a corollary of theorem 3.0.4.

We can also see that, if τ is merely n-setlike for some n, then $(n+2)$-stratified expressions are preserved.

Notice that we have not at this stage made the assumption that τ is a set. It is sufficient to assume that it is setlike. The distinction is important, for, as we shall see, there are sentences preserved by permutation models where the permutation involved is a *set* that are not always preserved when the permutation is merely setlike. It is natural to introduce the word 'invariant' to describe those Φ for which $\Phi \longleftrightarrow \Phi^\tau$ is T-provable for all (set) permutations τ. The way the word has been used so far applies to sentences preserved by all permutations that are *sets.* We will continue to use the word in this way, since we already have the word 'stratified' which, by theorem 3.0.4, is true of precisely the sentences preserved by setlike permutations. We should have a theory subscript for 'invariant', but it is usually safe to omit it. A theory T all of whose axioms can be T-proved to be invariant is said to be *invariant. NF* is invariant: thus, when τ is a permutation that is a set of V, V^τ is a model of *NF* iff V is. Evidently all theorems (of *NF*) are (*NF*-)invariant. We shall see some more results on invariant sentences in section 3.1.1.2. In particular, we shall learn that there are invariant formulae that are not equivalent to any stratified formula: we have seen that the axiom of counting is not equivalent to any stratified expression, and we shall see later that it is invariant.

We shall need to be explicit about the pairing functions we use, at least to the extent of knowing how to give a stratification to '$\langle x, y\rangle = z$'. We will in fact be using Quine ordered pairs here. These were introduced in section 2.1. Wiener–Kuratowski pairs would do too, of course, but Quine ordered pairs have the desirable feature that '$\langle x, y\rangle = z$' is stratified with 'x','y', and 'z' all being assigned the same subscript, so that, in '$y = f\text{`}x$', 'y' and 'x' are of the same type, and 'f' is of type one higher. This greatly simplifies the computations.

3.1.1 *Inner permutations in NF*

M^τ is a model of *NF* whenever M is, as long as τ is setlike. However, until section 3.1.2 we consider only those permutations that are also sets of the base model. This is because the simplest way of ensuring that a permutation is setlike is to stipulate that it be a set. In section 3.1.2 we will construct some independence proofs by means of outer permutations that are outer $\in$-automorphisms. This is the second simplest way of ensuring that a permutation is setlike!

By use of the permutation method we obtain new models which satisfy the same stratified sentences as the base model. If we want to understand what *un*stratified sentences are true in the resulting permutation model, we must use Henson's lemma. Now Henson's lemma does not apply straightforwardly to unstratified formulae, and to apply it to an unstratified formula we have to see the formula as a substitution instance of a stratified one. Suppose we are considering '$\exists x\Phi(x)^\tau$', where 'Φ' is weakly stratified but not stratified. That is to say, if we attempt to construct a stratification for "$\exists x\Phi(x)$" we will end up trying to give (at least, but for the sake of argument exactly) two different indices to 'x'. The proper thing to do here is to think of "$\Phi(x)$" as the result of substituting 'x' for 'y' in some formula '$\Psi(x, y)$' where 'x' and 'y' are of different types, so that the correct way to reveal the application of Henson's lemma to "$\exists x\Phi(x)^\tau$" is to think of 'x' as having two different types, so that some occurrences have 'τ_n' in front and some have 'τ_k'.

For the readers' peace of mind, the results in this section are presented in order of increasing difficulty rather than order of discovery. The first proof is really just a worked example.

PROPOSITION 3.1.3 Pétry–Hinnion. *Every model of NF has a permutation model with an $x = B`x$, a Boffa atom.*

Proof: a is a Boffa atom in $V^{(a,B`a)}$ iff $(\forall y)(y \in a \longleftrightarrow a \in y)^{(a,B`a)}$, that is $(\forall y)(y \in_{(a,B`a)} a \longleftrightarrow a \in_{(a,B`a)} y)$. $y \in_{(a,B`a)} a$ iff $y \in B`a$ iff $a \in y$. If $y \neq a$ and $y \neq B`a$, then $y$ is fixed, so $a \in y$ is the same as $a \in (a, B`a)`y$ which is $a \in_{(a,B`a)} y$ as desired. If $y = a$, then $a \in a$ is the same as $a \in B`a$ (by definition of $B$), which is $a \in (a, B`a)`a$ which is $a \in_{(a,B`a)} a$ as desired. If $y = B`a$ then $y \in_{(a,B`a)} a$ becomes $B`a \in B`a$ and $a \in (a, B`a)`y$ becomes $a \in a$, and these are equivalent as desired. Notice that since $a \in a$ iff $a \in B`a$, the Boffa atom is self-membered or not according to whether or not a is. ■

Adrian Mathias has noted that this apparently trivial observation shows that, if *NF* is consistent, the existence of antimorphisms (see section 1.2) of order 2 is independent of *NF*; this is because if a is a Boffa atom and π an antimorphism of order 2 a contradiction follows: we have $\pi`a \in a$ iff $\pi`a \in B`a$ iff $a \in \pi`a$ iff $\pi`a \notin \pi^2`a$ iff $\pi`a \notin a$. In fact, with a little bit of work we can prove the much stronger proposition 3.1.17 below.

PROPOSITION 3.1.4 *Every model of NF has a permutation model in which the universe is the union of two disjoint sets* I *and* II, *where* I $=$ $\mathcal{P}`$II *and* II $= b`$I.

Proof: It is actually not quite correct to use the suggestive 'I' and 'II' for this since we do not know that $x \in$ I$\rightarrow$I Wins G_x, but we can infer at least some of the consequences of $\in$-determinacy from it as we have seen. First we must reduce this problem. We seek x such that $x = \mathcal{P}`-x$ and $-x = b`x$. These two conditions are in fact equivalent (since $b`y = -\mathcal{P}` - y$), so it will be sufficient to find an x such that $x = \mathcal{P}` - x$. By Henson's lemma (and the remarks above), to construct τ so that V^τ contains such an x it will be sufficient to find x so that $\tau``x = \mathcal{P}` - x$. By proposition 2.1.10 it will be sufficient that $x \sim \mathcal{P}` - x$ and $-x \sim -\mathcal{P}` - x$. It is not hard to verify that, if $x$ is $B`w$ for some $w \notin w$, then both objects are of size $\overline{\overline{V}}$ (cf. appearance of $-B`\Lambda$ in the proof of proposition 2.1.15). ■

As announced, this is not full $\in$-determinacy, but it gives us some of the consequences of it as discussed on page 13.

PROPOSITION 3.1.5 *If NF is consistent, so is (for each k)* $NF + \bigwedge_{n \leq k}(\forall x)(x \neq \iota^n`x)$.

Proof: The permutation τ we need to prove the consistency of $\bigwedge_{n \leq k}(\forall x)(x \neq \iota^n`x)$ is the product of all transpositions $(\iota^{k+1}`x, -\iota^{k+1}`x)$. Suppose $(x = \iota^m`x)^\tau$ with $m \leq k$ which is $\tau`x = \iota\tau\iota\tau\iota\tau \ldots \iota\tau x$ with m *taus* on the right altogether.

There are two cases to consider.

(1) x fixed by τ. Let w be maximal so that $x \in \iota^w``V$. There is such a $w$ since $x \notin \iota^{k+1}``V$. Thus $x = \iota^w`y$ for some $y$ (possibly $w = 0$). Working from the inside out, $\tau$ does nothing except once it swaps something with its complement, so $x = \iota^{m-1-k+w}` - \iota^{k+1}``y$ and so, since $x \in \iota^w``V - \iota^{w+1}``V$, we have $w = m - 1 - k + w$ and $m = k + 1$, contradicting $m < k$.

(2) x is moved by τ. Now $\tau`x$ is a singleton, so $x$ is a complement of a singleton$^{k+1}$. Further we have $\iota^{k+1}`y = \iota\tau\iota\tau\iota \ldots \iota - \iota^{k+1}`y \ldots$ with $m - 1$ $\tau$s on the right hand side. Now as long as $m < k+1$, working from the inside as before, none of the occurrences of $\tau$ do anything; so $\iota^{k+1}`y = \iota^n` - \iota^{k+1}`y$ which is impossible as long as $n < k + 1$. ■

Set theorists with stouter hearts than mine may be prepared to write out the analogous proof that we can consistently assume that no set can be the (complement of the singleton)^n of itself. In this case, the permutation we need swaps with their complements objects that are (complement of the singleton)$^{n+1}$ of something.

PROPOSITION 3.1.6 Pétry [1974]. *Every model of NF has a permutation model with precisely n Quine atoms.*

Proof: By the preceding result, we can start in a model with no Quine atoms at all.[40] We saw at the outset that, in V^σ, x is a Quine atom iff

[40]There is some cheating here: all I am going to prove in this paragraph is that every model of *NF* has a permutation model that has a permutation model with precisely n Quine atoms. We shall see below that $(\exists\sigma(\exists\tau\Phi^\tau)^\sigma) \rightarrow \exists\tau\Phi^\tau$. This is theorem 3.1.20.

$\sigma`x = \iota`x$, so, if we want a model with precisely n Quine atoms, we simply swap n arbitrary objects with their singletons.

We may then use compactness to show that, if *NF* is consistent, it has models with infinitely many Quine atoms. However, we cannot prove this in *NF*, for the free n in the statement of proposition 3.1.6 is a metalanguage variable and, since every set of Quine atoms is strongly cantorian, an infinite such set would imply AxCount (see section 2.2) and thus the consistency of *NF*.

Pétry used this proof to obtain some of the results in section 2.1.3. If 'Φ' is stratified and $\mathcal{M}$ is a model with at least two Quine atoms then $\mathcal{M} \models \exists x \exists y(\phi(x,y) \longleftrightarrow \phi(y,x))$. If '$\phi$' is actually *homogeneous* then '$\exists x \exists y \phi(x,y) \longleftrightarrow \phi(y,x)$' is stratified and therefore must have been true in the base model. But any model has a permutation model with at least two Quine atoms, so $NF \vdash (\exists x)(\exists y)\phi(x,y) \longleftrightarrow \phi(y,x)$. ■

PROPOSITION 3.1.7 Henson [1973b]. *Every model of NF has a permutation model with a proper class of Quine atoms.*

Proof: Start in a base model with no Quine atoms at all. Consider the product π of all transpositions $(T\alpha, \iota`\alpha)$ for $\alpha$ an ordinal. We have seen earlier that if $\sigma$ swaps $x$ and $\iota`x$ then in V^σ x becomes a Quine atom. This, and the fact that we start with a model with no Quine atoms, ensures that something is a Quine atom in V^π iff it was an ordinal $\alpha = T\alpha$ in the base model. But, by the Burali-Forti paradox, these form a proper class in the base model, and therefore also in the permutation model. (If this class were coded by a set x in V^π, then in the base model the class would be the set $\pi^{-1}``x$.) ■

PROPOSITION 3.1.8 Hinnion–Mostowski. (see Hinnion [1975].) *Every model of NF has a permutation model in which Mostowski's collapse lemma holds for well-founded extensional relations on a strongly cantorian set.*

Proof: We need a notion of relational type of well-founded extensional relations with a maximal element as in remark 2.5.1. In short, isomorphism classes of relations that look like $\in$-graphs of transitive sets and we will resume notation from that section. I shall write "$bfext(R)$" to mean that R is a well-founded extensional relation with a top element, as we need here. This time we will need the extension of T to these objects, where T of a relational type α is simply the relational type of $RUSC`R$ for some $R \in \alpha$. Consider finally the product π of all transpositions $(T\alpha, \{\beta : \beta \; E \; \alpha\})$ for $\alpha \in BF$.

We want to know that V^π satisfies

$(\forall R)[(stcan(dom`R) \wedge bfext(R)) \rightarrow$
$(\exists y \exists f : dom`R \rightarrow y \subseteq \mathcal{P}`y \wedge (\forall z_1 z_2)(z_1 R \; z_2 \longleftrightarrow f`z_1 \in f`z_2))],$

that is

$(\forall R)[(stcan(dom`R) \wedge bfext(R)) \rightarrow$
$(\exists y \exists f : dom`R \rightarrow y \subseteq \mathcal{P}`y \wedge (\forall z_1 z_2)(z_1 R\ z_2 \longleftrightarrow f`z_1 \in f`z_2))]^{\pi}$

which is

$(\forall R)[(stcan(\pi_n`(dom`R)) \wedge bfext(\pi_n`R)) \rightarrow$
$\exists y \exists f : (\pi_n`dom`R) \rightarrow \pi`y \subseteq \mathcal{P}`\pi_2`y \wedge (\forall z_1 z_2)(z_1 R\ z_2 \longleftrightarrow f`z_1 \in f`\pi`z_2))]$.

Now what can this f and this y possibly be? If $stcan(dom`R)$ then R is going to be isomorphic to E restricted to the well-founded relational types hereditarily E the relational type of R. The y we want is simply this relational type of R, and f is the obvious homomorphism (which works when the domain of a relation is strongly cantorian) from the domain of R onto an initial segment of *BF*. This initial segment is simply the set of things that are $\in_\pi$ the relational type of R. ■

PROPOSITION 3.1.9 *Every model of NF has a permutation model with no Boffa atom.*

Proof: The proof here is reproduced from Forster [1983b] with a correction by Prof. M. Yasuhara.[41] Suppose not, so there is at least one Boffa atom. Let τ be the product of all transpositions $(B`x, -B`x)$. The assertion that (there is no Boffa atom)$^\tau$ is $\forall x \exists y (x \in \tau`y \longleftrightarrow y \notin \tau`x)$. We have two cases to consider: if $x = \tau`x$, then take $y$ to be a Boffa atom. If $x \neq \tau`x$, then either x is a value of B, in which case take y to be Λ, or it is $-B`z$ for some $z$, in which case take $y$ to be $\iota`z$, for then $\tau`y = y$. The left hand side of the biconditional is $x \in \iota`z$, that is $-B`z = z$, which is impossible (ask $z \in z$?). The right hand side is $\iota`z \notin B`z$, which is false. ■

PROPOSITION 3.1.10 *Every model of NF + AC_2 has a permutation model with a (non-trivial) $\in$-automorphism which is a set.*

Proof: The important fact here is that σ is an $\in$-automorphism iff it is a fixed point for j (exercise). So we consider the expression $(\sigma = j`\sigma)^\pi$. This is equivalent to $\pi_{n+1}`\sigma = j`(\pi_n`\sigma)$ for n sufficiently large by Henson's lemma. Now $\pi_{n+1}`\sigma$ is $(j^n`\pi)`(\pi_n`\sigma)$, so we are looking for a permutation π so that π and $j`\pi$ are conjugated by something in $J_n$ for some small concrete $n$. But if we can find $\pi$ so that $\pi$ and $j`\pi$ are conjugated by σ, say, then $j^n`\pi$ and $j^{n+1}`\pi$ are conjugated by $j^n`\sigma$. Therefore if we can find $\pi$ so that $\pi$ and $j`\pi$ are conjugates, we are done. Unfortunately, to find such a π we need some choice. Let *GC* (*Group Choice*: Forster [1987a]) be the axiom saying that sets of finite-or-countable sets have selection functions. Given that *AC* for sets of pairs is AC_2, this version of *AC* *ought* to be AC_ω, but this notation is usually used for *AC* for countable sets. Since

[41] By letter.

AC for sets of finite-or-countable sets seems to be needed to prove many desirable assertions of first-order group theory (e.g. that every element of a symmetric group is the product of two involutions,[42] or that in a symmetric group any two elements with the same cycle type are conjugate), I here use this mnemonic for it. GC looks innocent enough but it cannot be true in any model of NF where every set is definable by a stratified formula. This is because it implies AC_2, and therefore that there is a connected antisymmetric relation on V, and no such relation can be definable, as we saw in section 2.1.3. If we invoke GC, then we can show that two permutations with the same cycle type are conjugate; in fact, if we are content that the automorphism should be an involution, we will only need AC_2 instead of GC. It is easy enough to find an involution π so that both π and $j`\pi$ fix precisely $\overline{\overline{V}}$ things, and move precisely $\overline{\overline{V}}$ things and are therefore of the same cycle type. For example, let $\pi$ fix everything in $B`\Lambda$, and swap subsets of $\mathcal{P}` - \{\Lambda\}$ with their complements in $\mathcal{P}` - \{\Lambda\}$. Then $j`\pi$ fixes everything in $\mathcal{P}`(B`\Lambda)$ which is clearly of size $\overline{\overline{V}}$, and will move anything that is a union of a singleton from $\mathcal{P}` - \Lambda$ and a subset of $B`\Lambda$, and again there are $\overline{\overline{V}}$ such objects. ■[43]

In $NF + GC$, not only can we not prove there are no automorphisms that are sets, we cannot even prove that there is only a set of them.

THEOREM 3.1.11 *"$Aut(\langle V, \in \rangle)$ is a set" is not a theorem of any consistent stratified theory extending $NF + GC$.*

Proof: Let T be a consistent stratified theory extending $NF + GC$. Since T is a stratified theory, it does not prove AxCount. So let M be a model of $T + \neg$AxCount. We will devise (one cannot say "construct" since there will be some use of AC!) a permutation σ so that σ is of infinite order and has the same cycle type as $j`\sigma$.

Devising such a σ requires care. We shall want σ to have precisely $\overline{\overline{\iota``V}}$ n-cycles for each $n \leq \aleph_0$. Consider the set

$$S = \{\langle x, y, z \rangle : z \in V \wedge y < x \leq \aleph_0\}$$

(that is to say, $y \in \mathbb{N}$ and $x \in \mathbb{N} \cup \iota`\aleph_0$) and define a permutation σ^* of it by

> $\langle x, y, z \rangle \mapsto \langle x, (y + 1 \bmod x), z \rangle$ if x is finite (so that, if $x = 1$ $\langle x, y, z \rangle$ is a fixed point) and

[42] See Degen [1988] for a proof that this assertion is independent of ZF.

[43] Friederike Körner has pointed out to me that if we do not require the permutation to be internal (though it does have to be setlike) then we may be able to get by with assuming AC_2 only in the metalanguage.

$\langle x, y, z\rangle \mapsto \langle x, \pi`y, z\rangle$ if $x = \aleph_0$, where $\pi`(2n+2) = 2n$, $\pi`0 = 1$, and $\pi`(2n-1) = 2n+1$.

Now, σ^* is a permutation of S of infinite order, and S is obviously the same size as V. Further, for each n in $\mathbb{N} \cup \iota`\aleph_0$ we can find a set of $\overline{\overline{V}}$ objects belonging to distinct $n$-cycles. Therefore there is a similar permutation $\sigma$ of $V$. We now have to show that $j`\sigma$ also has this feature that for each n in $\mathbb{N} \cup \iota`\aleph_0$ we can find a set of $\overline{\overline{V}}$ objects belonging to distinct $n$-cycles. We must show that $j`\sigma$ has precisely $\overline{\overline{\iota``V}}$ $n$-cycles for each $n \leq \aleph_0$. Now, for any $n$, the set of $n$-cycles of $\sigma$ is a set of disjoint countable (at most) sets, and therefore by $GC$ has a selection set of size $\leq \overline{\overline{\iota``V}}$. So it will be sufficient to show that $j`\sigma$ has at least $\overline{\overline{\iota``V}}$ n-cycles for each $n \leq \aleph_0$.

Consider the set of n-cycles under σ, with $n \geq 2$. This is a collection of disjoint sets, and any two selection sets are objects belonging to distinct n-cycles of $j`\sigma$. So we must show there are at least $\overline{\overline{V}}$ selection sets. By means of GC, we can set up a commutative family of bijections between the cycles (by labelling one object in each cycle '0'), so that each object of order n can be thought of as a pair of a singleton and an integer $< n$. Thus any selection set can be thought of as a function assigning singletons to integers $< n$. Therefore there must be $n^{\overline{\overline{\iota``V}}}$ (which is to say $\overline{\overline{V}}$) such selection sets.

For the case $n = 1$, note that any set all of σ-fixed points is a $j`\sigma$-fixed point. But there are precisely $\overline{\overline{V}}$ $\sigma$-fixed points, so there are precisely $\overline{\overline{V}}$ $j`\sigma$-fixed points as desired.

We now invoke GC to infer the existence of a τ that conjugates σ and $j`\sigma$. Recall that $M \models T + \neg$AxCount. In M^τ σ is an automorphism of infinite order. Let G be $Aut(\langle M, \in\rangle)$ and suppose that G is a set in M^τ. Now we can always prove by induction on n that, for any element a of any group H, a^n is also in H. This obviously depends on H being a set! We have assumed that G is a set, and we know that it contains an element of infinite order, so it is infinite. But it is a simple matter to show that if G is a set, it is strongly cantorian. (ι restricted to G is simply $\{\langle \iota`\sigma, j`\sigma\rangle : \sigma \in G\}$.) Since G is infinite and strongly cantorian, we infer AxCount by proposition 2.1.3. But AxCount is invariant (theorem 3.1.26) so it was true in M, contradicting the hypothesis. Therefore G is not a set in M^τ, so T cannot prove that G is a set. ∎

Both these proofs use permutation models that construct automorphisms one cycle at a time. If AxCount holds, the cycle may be infinite, but they do not suffice to show that $Aut(\langle V, \in\rangle)$ need not be cyclic. It remains an open problem to arrange for the group of inner automorphisms not to be cyclic. In general, very little seems to be known about the group theory of $Aut(\langle V, \in \rangle)$. Any model of *NF* with a non-trivial inner automorphism σ violates strong extensionality (σ gives a winning strategy to player $=$), so it would be nice to be able to prove the consistency of "V is rigid".

THEOREM 3.1.12 *Every consistent Σ_1^{Levy} sentence is consistent with NFC. More precisely,*

$$NFC \vdash (Con(Ext + \phi) \to (Con(NFC) \to Con(NFC + \phi)))$$

for $\phi \in \Sigma_1^{Levy}$.

Proof: Let Φ be a consistent Σ_1^{Levy} sentence. By the Skolem–Löwenheim theorem the theory Φ plus extensionality has a model in the integers, $\langle \mathbb{N}, R \rangle$. What we would like to be able to do is swap each integer n with R^{-1}"ι'n, which would give us a transitive copy of $\langle \mathbb{N}, R \rangle$. Unfortunately we cannot do this in *NF*, since there is no guarantee that this permutation will be a set, the defining condition being unstratified. But it will be a set if we have the axiom of counting, so we can do it in *NFC*. This tells us that there is a bijection $f : \iota$"$\mathbb{N} \to \mathbb{N}$ which is in fact ι^{-1}. Now we let σ be the product of all transpositions (f'ι'n, R^{-1}"ι'n) for all $n \in \mathbb{N}$. We have to check that in the permutation model $\langle \mathbb{N}, \in_\sigma \rangle$ is a transitive copy of $\langle \mathbb{N}, R \rangle$. Suppose $n \in_\sigma m$. Then $n \in R^{-1}$"$(f \circ \iota)^{-1}$'m so $n \; R \; (f \circ \iota)^{-1}$'$m$. But $(f \circ \iota)^{-1}$'$m = m$, so it is an isomorphic copy as desired. For transitivity we need verify only that anything $\in_\sigma$ a natural number is another natural number and this is immediate from construction.

Φ is true in $\langle \mathbb{N}, R \rangle$, in $\langle \mathbb{N}, \in_\sigma \rangle$ by the above, and is therefore true in the permutation model V^σ since Σ_1^{Levy} sentences are preserved under end-extension. The permutation model is still a model of the axiom of counting because that axiom is invariant (see theorem 3.1.26). ■

This is in fact the best possible. We cannot use *NF* instead of *NFC* here, for AxCount is Σ_1^{Levy} and independent of *NF*.

It is worth reflecting on what would happen if we tried to prove something similar for stratified Σ_1^{Levy} sentences in *NF*. As before, there is a bijection $f : \iota$"$\mathbb{N} \to \mathbb{N}$, but this time we do not know that f is ι^{-1}, so we have to be careful. Again σ is the map that swaps f'ι'n with R^{-1}"ι'n for all $n \in \mathbb{N}$, and leaves everything else alone. Everything is the same as the preceding case until we reach the line $n \; R \; (f \circ \iota)^{-1}$'$m$. This time we do not know that $(f \circ \iota)^{-1}$ is the identity, so all we have shown is that $\langle \mathbb{N}, \in_\sigma \rangle$ is a transitive copy of a permutation model of $\langle \mathbb{N}, R \rangle$. We would like to be able to appeal to the fact that since Φ is stratified it is preserved under permutation models, so that $\langle \mathbb{N}, \in_\sigma \rangle$ is a transitive model of Φ as desired. Therefore we could conclude that the ground model satisfied Φ. But the ground model was arbitrary, so Φ is a theorem of *NF*.

What scuppers this is that the permutation $(f \circ \iota)^{-1}$ is not a set of the model. Worse, there seems no reason to suppose that it is setlike. It can be shown to be 1-setlike, but all that gives us is a proof that if ϕ is Σ_1^{Levy} and 3-stratified, and *NF* proves that the theory ext + ϕ has a model, then $NF_3 + \phi$ has a model. But we know results like this from proposition 2.3.14 anyway.

THEOREM 3.1.13 *Every model $\mathcal{M}$ of NF has a proper end-extension M' which satisfies the same stratified sentences as $\mathcal{M}$ and for which the inclusion embedding preserves power set.*

Proof: Let π be a stratified (not necessarily homogeneous) *definable* 1-1 (but not surjective) function so that $(\forall xy)(\pi`x = \pi``y \to x = y)$. Let $\sigma$ be the product of all transpositions $(\pi`x, \pi``x)$. (We need π to be definable so that the definition of σ is stratified.) It will turn out that, in V^σ, $\pi`V$ is an isomorphic copy of the old V.

Suppose $x \in_\sigma y \in_\sigma \pi`V$. Now $y \in_\sigma \pi`V$, so $y \in \pi``V$, and so $y$ is $\pi`z$ for some $z$. Also, $x \in_\sigma \pi`z$ iff $x \in \pi``z$, so $x = \pi`w$ for some w. Further, $x \in \pi``V$ iff $x \in_\sigma \pi`V$, so $\pi`V$ is transitive$^\sigma$.

Suppose $(x \subseteq \pi`V)^\sigma$. Then $\sigma`x \subseteq \sigma`\pi`V = \pi``V$, so $\sigma`x = \pi``y$ for some y so $x = \pi`y$. Therefore $x \in \pi``V$ and $x \in_\sigma \pi`V$. So $\pi`V$ is transitive$^\sigma$ and (extends its own power set)$^\sigma$, and is therefore (equal to its own power set)$^\sigma$. It remains to be shown that it is an isomorphic copy of the old V.

Suppose $x \in_\sigma y \in_\sigma \pi`V$. As before, $y \in \pi``V$ so $y$ is $\pi`z$ for some $z$. If $x \in_\sigma \pi`z$, then $x \in \pi``z$, so $x = \pi`w$ for some $w$. Now $\pi`w \in_\sigma \pi`z$, so $\pi`w \in \pi``z$ and $w \in z$ which is to say that π^{-1} is an isomorphism between $\langle \pi`V, \in_\sigma \rangle$ and $\langle V, \in \rangle$, so $\langle V, \in_\sigma \rangle$ is an end-extension of $\langle V, \in \rangle$ as desired.

All we need now is to find such a π. In fact there are many: B is one; for each n, ι^n is another. One such was used by Henson in [1973b]. All these are inhomogeneous. There are actually lots of homogeneous maps too. All we need is a definable total map whose range is disjoint from its power set. If a is some object not a member of itself, then $B`a$ is an object disjoint from its power set (remark 2.1.12) and of the same size as $V$, so, if $a$ is a closed term, there will be a definable bijection $V \longleftrightarrow B`a$. The cardinal number 0 ($= \iota`\Lambda$) is such an a, and the function $\lambda x.(x - \mathbb{N} \cup S``(x \cap \mathbb{N}) \cup \iota`0)$ (whose range is $B`0$ and which is one of the functions used in the definition of Quine ordered pairs) is such a π. Homogeneous maps like this presumably give rise to permutation models in which the old universe appears in the new as a set the same size as the (new) universe. ■

REMARK 3.1.14 *If consistent, NF has no $\Pi_2^{\mathcal{P}}$ axiomatization.*

Proof: Iterate the construction of theorem 3.1.13, and take a direct limit. It satisfies all $\Pi_2^{\mathcal{P}}$ sentences true in the original model, but does not contain a universal set. ■

The proof of theorem 3.1.13 is a refinement of the simple observation that if π is a permutation that maps $x \longleftrightarrow \mathcal{P}`x$ then, in $V^{\pi}$, $\pi^{-1}`x$ is identical to its own power set. In this case we have chosen an x which is (seen from outside) a 1-1 copy of V and therefore the permutation model is (isomorphic to) an end-extension of the original model. If we choose π carefully, then the new model contains no $x = \mathcal{P}`x$ strictly between the old model and the new. This construction is fairly robust: the product $\sigma$ of all transpositions $(\iota`x, \iota``x)$ can be used in *ZF* to give a model that is (isomorphic to) an end-extension of the old model. It will not be a well-founded end-extension of the model we start with, because $\Lambda = \pi``\Lambda$ and so σ will swap Λ and $\iota`\Lambda$ making Λ a Quine atom in the extension, and of course the same happens here.

Permutations giving us models containing sets identical to their power sets lead naturally to a theorem of Boffa on typed properties. Recall that an expression $\Phi(x)$ is *typed* if there is a closed stratified Ψ so that $\Phi(x)$ is the result of replacing all quantifiers Qy in Ψ by the restricted quantifier $Qy \in \mathcal{P}^{n}`x$, where n is the type 'y' receives in some fixed stratification. It is not hard to see that $\overline{\overline{x}} = \overline{\overline{y}}$ implies that x and y have the same typed properties. This is because $\langle\langle x\rangle\rangle \simeq \langle\langle y\rangle\rangle$ in these circumstances.

PROPOSITION 3.1.15 Boffa [1975a]. *Let $\Phi(\ )$ be a typed property so that NF $\vdash \Phi(V)$. Then NF $\vdash (\forall x)(\overline{\overline{x}} = \overline{\overline{\mathcal{P}`x}} \rightarrow \Phi(x))$.*

Suppose $\overline{\overline{x}} = \overline{\overline{\mathcal{P}`x}}$. Then either $\overline{\overline{x}} = \overline{\overline{V}}$, in which case the result is immediate, or it is smaller. If $\overline{\overline{x}} < \overline{\overline{V}}$, Bernstein's lemma tells us that, since $\overline{\overline{-x}} + \overline{\overline{x}} = \overline{\overline{V}} \cdot \overline{\overline{V}}$ then $\overline{\overline{-x}} \geq \overline{\overline{V}}$ (which is what we want) or $\overline{\overline{x}} \geq_* \overline{\overline{V}}$ (which we must improve). If the latter, then $\overline{\overline{x}} = \overline{\overline{\mathcal{P}`x}} \geq \overline{\overline{\mathcal{P}`V}} = \overline{\overline{V}}$, contradicting hypothesis. So $\overline{\overline{-y}} = \overline{\overline{V}}$ holds for any y such that $\overline{\overline{y}} = \overline{\overline{\mathcal{P}`y}} \wedge \overline{\overline{y}} < \overline{\overline{V}}$. In particular, it holds for $x$, $\mathcal{P}`x$, $\mathcal{P}^{2}`x$, $\ldots$ . This means we can invoke proposition 2.1.10, to conclude that there is a permutation $\pi$ of $V$ mapping $x$ onto $\mathcal{P}`x$. Then $(y = \mathcal{P}`y)^{\pi}$ is $(\forall z)(\pi`z \subseteq \pi`y \longleftrightarrow \pi`z \in \pi_2`y)$. This is $\mathcal{P}`(\pi`y) = \pi``\pi`y$, which is certainly true for $y = \pi^{-1}`x$. So, in V^{π}, $\pi^{-1}`x$ is identical to its power set and is accordingly a model of *NF*. Therefore $\Phi(\pi^{-1}`x)^{\pi}$. Now $\Phi(y)^{\pi}$ is $\Phi(\pi`y)$, since any typed $\Phi(\ )$ is a 1-formula. So $\Phi(x)$ as desired. ■

There are other proofs of this important result, and some elaborations of it. Pétry makes the following observation in [1979]. Although we can infer $\Phi(x)$ for typed Φ if we know $\overline{\overline{x}} = \overline{\overline{\mathcal{P}`x}}$, there is no converse, for otherwise, for some typed Φ, '$\alpha = 2^{T\alpha}$' will be equivalent to '$\exists x \in \alpha \wedge \Phi(x)$' (which is stratified) so its extension will be a *set* of cardinals. But by consideration of Hartogs' aleph function, it is easy to show that the collection of such α

cannot be a set. (*Hint:* consider the least aleph that is a Hartogs' aleph of some member of this set.) Therefore, for any typed Φ, there will always be many x such that $\Phi(x)$ with x and $\mathcal{P}`x$ of different sizes.

PROPOSITION 3.1.16 *If $AxCount_{\leq}$ fails, then there is a permutation model in which there is a finite set $X \supseteq$ all well-founded sets.*

Proof: If $\text{AxCount}_{\leq}$ fails there is an $n \in \mathbb{N}$ such that $2^{Tn} < n$. Remark 2.1.12 tells us that every cardinal contains a set disjoint from its power set, and this makes it easy, for this n, to obtain an x and a permutation π such that $\overline{\overline{x}} = n \wedge \pi``\mathcal{P}`x \subseteq x$ and, in $V^\pi$, $\mathcal{P}`\pi`x \subseteq \pi`x$. ■

PROPOSITION 3.1.17 *The duality scheme "$\phi \longleftrightarrow \hat{\phi}$" of section 1.2 is not a theorem scheme of NF.*

Proof: Recall that $\hat{\phi}$ is the result of replacing all occurrences of $\in$ in ϕ by $\notin$ and vice versa. A Quine *antiatom* for the nonce is a set $x = -\iota`x$. The hat of "($\exists$ Quine atom)" is "($\exists$ Quine antiatom)" so, if the scheme $\phi \longleftrightarrow \hat{\phi}$ is provable, then any model of *NF* with either a Quine atom or antiatom must have one of the other as well. Suppose that the proposition is false and that we have a model of *NF* with no Quine atoms, and therefore no Quine antiatoms. Consider the permutation model arising from the transposition $(a, \iota`a)$ for some arbitrary object $a$. In the permutation model, $a$ is a Quine atom, so the permutation model must contain a Quine antiatom too: $b$, say. Evidently $b \neq a$, so $(a, \iota`a)`b = b$, and it is easy to show that b was a Quine antiatom in the model we started with, contradicting duality. ■

Something slightly weaker than the duality scheme *is* provable.

PROPOSITION 3.1.18 *$NF \vdash \Phi$ iff $NF \vdash \hat{\Phi}$.*

Proof: $\hat{\Phi}$ is the special case of Φ^σ when σ is complementation. Therefore any model of $\neg\Phi$ gives rise to a model of $\neg\hat{\Phi}$. But the hat operation is an involution, so one is provable iff the other is. ■

There are a number of sentences that give the impression that their consistency should be fairly easily provable by permutations, since they must be true in all term models or models in which every set is symmetric, but for which no such proofs have yet been found. We conclude with a list of some.

"There are no non-trivial $\in$-automorphisms of the universe." Since there is an automorphism in V^σ iff $(\exists\pi)(\sigma\pi\sigma^{-1} = j`\pi)$ it will be sufficient to find $\sigma$ such that $(\forall\pi)(\pi \neq 1 \rightarrow \sigma\pi\sigma^{-1} \neq j`\pi)$.
"Every self-membered set is infinite." To prove this relative to *NF*, it would be sufficient to find a π for which we can prove that $x \in \pi`x$ implies that $\pi`x$ is infinite. Körner [1994] has shown that it is consistent relative to *NF*

that there should be such a permutation, but it remains an open problem to prove the existence of one in *NF*.
"Every self-membered set has a member that is not self-membered."
"No self-membered set belongs to all its members."

The last three are $\forall^*\exists^*$, and the second is a $\forall^*\exists^*$ scheme. It is relatively easy to show the consistency of $\exists^*\forall^*$ sentences when possible at all, and permutations with finite support have always been sufficient, and these can be constructed by hand. The construction of permutations giving us $\forall^*\exists^*$ sentences tends to be much harder, and always seems to need permutations with infinite support. At the moment we know of no general results of the form "All formulae in Γ that are consistent can be proved consistent by means of permutations." We are sorely in need of a topology or some other structure on J_0 that would enable us to give a (presumably non-constructive) proof of something of this kind. The difficulty is (Gaughan [1967]) that a topology on J_0 is Hausdorff iff all stabilizers of singletons are open, iff they are all closed. Thus the collection of permutations giving rise to Quine atoms is liable to be open dense, which is not the flavour we want.

REMARK 3.1.19 *Let $\mathcal{T}$ be the topology whose basic neighbourhoods are boolean combinations of sets of the form $\{\sigma : x \in \sigma`y\}$:*
(i) *$\mathcal{T}$ is Hausdorff;*
(ii) *$\mathcal{T}$ is incompact;*
(iii) *J_0 acts $\mathcal{T}$-continuously on itself by skew-conjugation.*[44]

Proof: (i) If σ and τ belong to the same basic open sets, then $(\forall x)(\forall y)(x \in \sigma`y \longleftrightarrow x \in \tau`y)$, which is to say $(\forall y)(\sigma`y = \tau`y)$, that is $\sigma = \tau$.

(ii) Fix x. Then $\{\{\tau : x \in \tau`y\} : y \in V\}$ is an open cover of J_0 with no finite subcover. I am indebted to Tony Maciocia for pointing this out to me.

(iii) Let $O_{x,y}$ be the basic open set $\{\tau : x \in \tau`y\}$ and let $\sigma$ be an arbitrary (set) permutation. When $\sigma$ acts by skew-conjugation, it maps $O_{x,y}$ to $\{(j`\sigma)\tau(\sigma^{-1}) : x \in \tau`y\}$, which is $\{\tau : \sigma`x \in \tau\sigma`y\}$, which is $O_{\sigma`x,\sigma`y}$. ■

3.1.1.1 *The modal logic* The alert reader may have noticed in the previous section that various claims along the lines "every model of *NF* has a permutation model in which ..." were made when all that was proved was that every model in which something specific happens satisfies ... and the something that happens can be arranged by a permutation. In other words, we will need to know

[44] See page 115 for a definition of skew-conjugation.

$$\exists\pi(\exists\sigma(\Phi^\sigma)^\pi) \to \exists\sigma(\Phi^\sigma).$$

Slaves to the deplorable modern habit of seeing modal operators everywhere will immediately recognize this as a version of the S4 axiom. As it happens, this is one of the very few totally unproblematic uses of modal logic known.[45] Here we characterize the modal logic that arises from reading "$\forall\sigma\Phi^\sigma$" as "$\Box\Phi$". Since this commits us to quantifying over permutations we must here restrict ourselves to permutations that are *sets*. We have in fact been doing this but, until now, this assumption has not been critical. The modal logic this gives rise to is S5 + Barcan + converse Barcan + Fine's principle H which we will explain below.

THEOREM 3.1.20 *The propositional part of the logic is precisely* S5.

Any proof in *NF* of a wff ϕ translates uniformly to a proof of ϕ^τ *with* τ *free.* This shows that the rule of necessitation holds. $\Box$ clearly distributes over $\to$ since it is a secret $\forall$; $\Box p \to p$ follows similarly. All that remains is the K5 axiom $\Diamond\Box p \to \Box p$. To verify this, it will be sufficient to show the following.

LEMMA 3.1.21 $\Box\phi(\vec{x})$ *is invariant.*

$\Box\phi(\vec{x})^\tau$ is $(\forall\sigma)(\phi(\vec{x})^\sigma)^\tau$, which is $(\forall\sigma \in J_0)(\phi(\vec{x})^\sigma)^\tau$ which is $(\forall\sigma \in J_0)(\tau_n`\sigma \to (\phi(\vec{x})^\sigma)^\tau)$ for $n$ sufficiently large. What is $(\phi(\vec{x})^\sigma)^\tau$? Any atomic formula '$u \in v$' in $\Phi$ becomes ' $(u \in \sigma`v)^\tau$ '. Now ' $(u \in \sigma`v)$ ' is just another stratified formula with three free variables so we can apply lemma 3.1.2 (Henson's lemma) to ' $(u \in \sigma`v)^\tau$ ', getting ' $\tau_n`u \in (\tau_{n+2}`\sigma)\tau_{n+1}`v$ '. Here the '2' in the subscript '$n + 2$' occurs because we are using Quine ordered pairs. Had we been using Wiener–Kuratowski ordered pairs, it would have been '4', and the proof would have been greatly complicated. Now $\pi`u \in v \longleftrightarrow u \in (j`\pi)^{-1}`v$, and $(j`\tau_n)^{-1}$ is $\tau(\tau_{n+1})^{-1}$, so this is equivalent to $u \in \tau(\tau_{n+1})^{-1}(\tau_{n+2}`\sigma)\tau_{n+1}`v$. Also $\tau_{n+2}`\sigma$ is $\tau_{n+1}\sigma(\tau_{n+1})^{-1}$, so it reduces to $u \in \tau\sigma`v$. Now multiplication on the left by $\tau$ is simply a permutation of the symmetric group on $V$ so, since the $\sigma$ is quantified, we can reletter it to get $u \in \sigma`v$. So we conclude that $\Box\phi(\vec{x})^\tau$ is equivalent to $\Box\phi(\vec{x})$.

Thus the propositional part of the logic is at least S5. By proposition 3.1.6, we know that every model of *NF* has a permutation model with precisely n Quine atoms, so there are infinitely many non-elementarily-equivalent possible worlds and a theorem of Scroggs [1951] says that in these circumstances if the logic is at least S5 then it is precisely S5. ∎

[45]The others are Solovay's interpretation of a logic G and a variation due to Goldblatt (see Boolos [1979]).

The validity of Barcan and converse Barcan ("$\forall x \Box \Phi(x) \longleftrightarrow \Box \forall x \Phi(x)$") under this interpretation follows simply from the fact that like quantifiers commute. The only quantificational principle that involves any hard work is *Fine's principle H* (Fine [1978]). H is the following:

$$(\forall \vec{x})(\forall \vec{y})(\text{Diff}(\vec{x}) \wedge \text{Diff}(\vec{y}) \to (\Box\Phi(\vec{x}) \longleftrightarrow \Box\Phi(\vec{y})))$$

where $\text{Diff}(\vec{x})$ is the conjunction of all inequations between the $\vec{x}$. The clauses with "Diff" are needed because otherwise we could falsify H trivially by taking as Φ some open formula that compelled some arguments to be distinct, or to be identical.

LEMMA 3.1.22

$$\exists\vec{x}(\text{Diff}(\vec{x}) \wedge \Box\phi(\vec{x})) \quad \to \quad \forall\vec{x}(\text{Diff}(\vec{x}) \to \phi(\vec{x})).$$

Proof: Let us suppose the antecedent true and let $\vec{a}$ be a set of witnesses to it. Let $\vec{y}$ be an arbitrary tuple (of the right length) of distinct things so that $\text{diff}(\vec{y})$. Let τ be some arbitrary permutation sending each a_i to y_i. Since $\vec{a}$ and $\vec{y}$ are sequences without repetitions, this will be possible. We do not care what τ does to the $\vec{y}$ (only what it does to the $\vec{a}$), and in any case it might not be possible to take it to be the product of the transpositions ¡(a_i, y_i) since $\vec{y}$ and $\vec{a}$ may overlap.

Let σ be $(j`\tau)^{-1}\tau$. Since $\Box\phi(\vec{a})$ we conclude, in particular, that $\phi(\vec{a})^\sigma$. A typical atomic formula $x \in y$ in $\phi$ becomes $x \in \sigma`y$ in $\phi(\vec{a})^\sigma$. Now $x \in \sigma`y$ is $x \in (j`\tau)^{-1}\tau`y$. In general $(\forall\pi)(\pi`x \in y \longleftrightarrow x \in (j`\pi)^{-1}`y)$, so this is $\tau`x \in \tau`y$. Now every variable and constant has a τ in front of it: the τ disappear from the bound variables as before, and the $\vec{a}$ are turned into $\vec{y}$. Thus $\phi(\vec{a})^\sigma$ is simply $\phi(\vec{y})$. ■

We can now prove H.

THEOREM 3.1.23

$$\exists\vec{x}(\text{Diff}(\vec{x}) \wedge \Box\phi(\vec{x})) \quad \to \quad \forall\vec{x}(\text{Diff}(\vec{x}) \to \Box\phi(\vec{x})).$$

Let $\vec{a}$ be a witness to the antecedent as before, so $\Box\phi(\vec{a})$. Then, by lemma 3.1.21, $\Box\Box\phi(\vec{a})$. Now apply lemma 3.1.22 with $\Box\phi(\vec{a})$ instead of $\phi(\vec{a})$. ■

It will be sometimes useful to think of invariance modally: thus ϕ is T-invariant iff $T \vdash \Diamond\phi \longleftrightarrow \Box\phi$.

It is worth noticing that, although the proofs in this section do depend on the quantifier implicit in the '$\Box$' being universal over permutations that are *sets*, the only proof that appears to need that quantifier to range over *all* permutations is that of $\Box p \to p$. It seems probable that the remaining proofs would work equally well if we restricted ourselves to permutations with finite support or, probably, any other normal subgroup of J_0 (in which case $\Box p \to p$ would still hold), or used any other suitably closed monotone quantifier such as the cofinite quantifier. Thus, although the matter has never been specifically investigated, we can expect that such structures would be models for K5 + H + Barcan + converse Barcan.

3.1.1.2 *Classes of invariant formulae* As we have seen, the sentences preserved by arbitrary setlike permutations are precisely the stratified sentences. Although it is a simple matter to show that in contrast the class of T-invariant sentences is not recursive for any sensible T, there are some natural classes of sentences which, although not stratified, are preserved by permutations that are *sets* and are therefore invariant. We may as well have a proof of this non-recursiveness before we proceed to examples of such natural classes.

REMARK 3.1.24 *If NF is consistent, there is a Φ such that NF does not decide whether or not Φ is (NF-) invariant.*

Proof: Let Φ be $(\exists x = \iota`x) \vee R$, where R is some arithmetic expression unprovable in *NF* and irrefutable in *NF* (as long as *NF* is consistent)[46] such as a Rosser sentence. Certainly $NF \vdash \Diamond\Phi$, since provably there are permutations making the first disjunct true. $NF \vdash \Box\Phi$ is not possible, however. We argue by *reductio ad absurdum.* Let M be an arbitrary model of *NF*. It has a permutation model in which there are no Quine atoms but in which (because by hypothesis $NF \vdash \Box\Phi$) Φ is true. Therefore R is true in that model. But R is invariant (see theorem 3.1.26 below) and so must be true in M. But M was arbitrary, so $NF \vdash R$ which we know is not the case. So *NF* does not prove that Φ is invariant. On the other hand *NF* cannot prove that Φ is not invariant either, for to do that, it would have to be a theorem of *NF* that $(\exists\sigma)(\Phi \longleftrightarrow \neg\Phi^\sigma)$. Placing ourselves in an arbitrary model M we can argue that since Φ is false in either M or M^σ then R similarly must be false in either M or M^σ. Now we know that R is invariant, so it must be false in both, in particular false in M. But M was arbitrary, so $NF \vdash \neg R$, contradicting the fact that R is consistent with *NF*. ■

COROLLARY 3.1.25 *If T is a consistent recursively axiomatizable theory extending NF, then the set of T-invariant sentences is not recursive.*

We have seen one example of a natural class of invariant sentences (the modalized sentences) in the previous section. In this section we will see that (roughly) any expression of cardinal arithmetic, or, more generally,

[46] Maurice Boffa suggested the possibility of a proof along these lines.

the theory of relational types, will be invariant. We will also prove a result related to a remark of Pétry's [1976] to the effect that any sentence which fails to be stratified only because of some occurrences of "$can(x)$" or "$stcan(y)$" is nevertheless invariant.

Pétry's remark that '$can(x)$' behaves as if it were stratified is actually a special case of something more general.

THEOREM 3.1.26 *Let $\phi(x)$ be the result of substituting 'x' for 'y' in '$\psi(x,y)$' which is stratified but inhomogeneous. Suppose also that $(\forall xyy')(\psi(x,y) \wedge \overline{\overline{y}} = \overline{\overline{y'}} \rightarrow \psi(x,y'))$. Then any sentence which fails to be stratified only because it contains occurrences of ψ is nevertheless invariant.*

Any such weakly stratified $\phi(x)$ is equivalent to $\exists y(\psi(x,y) \wedge y = x)$ for some stratified ψ, with 'x' of type n, 'y_i' of type m, and $m > n$. Also $\phi(x)^\sigma$ is $\exists y(\psi(x,y) \wedge y = x)^\sigma$, which is $\psi(\sigma_n\text{‘}x, \sigma_m\text{‘}x)$ by the usual transformations. But now, since $\overline{\overline{\sigma_m\text{‘}x}} = \overline{\overline{\sigma_n\text{‘}x}}$ (at least if $n \geq 1$), we conclude that $\phi(x)^\sigma$ is equivalent to $\psi(\sigma_n\text{‘}x, \sigma_n\text{‘}x)$—which is $\phi(\sigma_n\text{‘}x)$ (as long as $n \geq 2$), so $\phi(\ )$ behaves as if it were an n-formula. ■

In particular, all of cardinal arithmetic is invariant. All we now need to obtain Pétry's remark is the fact that '$can(x)$' is an expression of this form. In fact $\psi(\ ,\ )$ is just '$\overline{\overline{x}} = \overline{\overline{\iota\text{“}y}}$'. No doubt there are versions of this theorem for $\phi(x)$ where the 'x' receives more than two indices.

The fact that the same holds for "$stcan$" as well as "can" does not actually follow from the above, but it is easy enough to check that $stcan(x)^\pi \longleftrightarrow stcan(\pi\text{‘}x)$, which will enable us to prove Pétry's remark for $stcan$.

Forster's [1983a] result on the invariance of the theory of relational types is only sketched there. Here is a more precise statement and a proper proof.

THEOREM 3.1.27 *Every expression that is the result of binding all variables in some weakly stratified expression where all the free variables are restricted to being relational types or sets-of ^{n} relational types is invariant.*

Proof: We will need the fact that all relational type are k-symmetric for some fairly small k, and that sets-of ^{n} relational types are $(k+n)$-symmetric. Consider such a $\phi(\vec{x})$, where we think of some of the x_i as occurring several times, possibly with different type assignments at its several occurrences. Then $(\phi(\vec{x}))^\sigma$ becomes ϕ with each occurrence of a free variable having a prefix σ_m, where all the various m can be made sufficiently large. But now, since all variables range over relational types, or sets-of^n relational types and these are $(k+n)$-symmetric, we can reletter some of the $\sigma_m\text{‘}x$ judiciously (as long as $m \geq k+n$) so that every occurrence of x_i has the same prefix. Thus ϕ will behave as if it had a stratification assigning k to variables ranging over relational types, and $k+n$ to variables ranging over sets-of^n relational types. ■

We can now complete the treatment of AxCount$_\leq$ foreshadowed in section 2.1.1, theorem 2.1.4. The major result is as follows.

THEOREM 3.1.28 *The following are equivalent:*

1. *AxCount*$_\leq$
2. $\Diamond$ ($\bigcap\{y : (\Lambda \in y) \wedge (\mathcal{P}``y \subseteq y)\}$ *is a set*).

Proof:

$2 \to 1$

Since the conclusion of this conditional is invariant, it will be sufficient to deduce it from the existence of $\bigcap\{y : (\Lambda \in y) \wedge (\mathcal{P}``y \subseteq y)\}$. So assume the collection

$$\bigcap\{y : (\Lambda \in y) \wedge (\mathcal{P}``y \subseteq y)\}$$

is a set. We'd better have a name for it, $\mathtt{X}$, say. We are going to deduce AxCount$_\leq$.

First we show that $\mathtt{X}$ is a well-founded set. Suppose $z \in \mathtt{X}$. $\mathcal{P}`x \subseteq x$. Then $\mathcal{P}``(\mathcal{P}`x) \subseteq \mathcal{P}`x$ and $\Lambda \in \mathcal{P}`x$ so $z \in \mathcal{P}`x$. But $\mathcal{P}`x \subseteq x$ so $z \in x$ as desired.

Next we show that $\mathtt{X}$ is totally ordered by $\subseteq$. Let x be $\in$-minimal such that $(\exists y)(x \not\subseteq y \not\subseteq x)$, and let y be $\in$-minimal such that $x \not\subseteq y \not\subseteq x$. In fact we can take these to be power sets and so we have x and y such that $x \subseteq y \vee y \subseteq x$ (by $\in$-minimality) but $\mathcal{P}`x \not\subseteq \mathcal{P}`y \not\subseteq \mathcal{P}`x$ which is clearly impossible.

Since $\mathtt{X}$ is totally ordered by $\subseteq$ we must have $(\forall x)(x \subseteq \mathcal{P}`x \vee \mathcal{P}`x \subseteq x)$. The second disjunct contradicts foundation so we must have $(\forall x \in \mathtt{X})(x \subseteq \mathcal{P}`x)$.

Next we prove by induction that each member of $\mathtt{X}$ is finite (has cardinal in $\mathbb{N}$). Suppose not, and let $\mathcal{P}`x$ be a $\in$-minimal infinite member of $\mathtt{X}$. But if $\overline{\overline{\mathcal{P}`x}} \notin \mathbb{N}$ then clearly $\overline{\overline{x}} \notin \mathbb{N}$ too.

Notice also that there can be no $\subseteq$-maximal member of $\mathtt{X}$, for if x were one we would have $\mathcal{P}`x \subseteq x$ and $x \in x$ contradicting foundation.

Therefore the sizes of elements of $\mathtt{X}$ are unbounded in $\mathbb{N}$. Now let n be an arbitrary member of $\mathbb{N}$. By unboundedness we infer that for some $x \in \mathtt{X}$ we have $\overline{\overline{x}} \leq n \leq \overline{\overline{\mathcal{P}`x}}$ and therefore $\overline{\overline{x}} \leq n \leq \overline{\overline{\mathcal{P}`x}} \leq 2^{Tn}$. But n was arbitrary, so $(\forall n \in \mathbb{N})(n \leq 2^{Tn})$. But by theorem 2.1.4 this is equivalent to AxCount$_\leq$.

$1 \to 2$

The relation E on natural numbers is defined by $x \; E \; y$ iff the xth bit of the binary representation of y is 1. The permutation we want is

$$\pi = \prod_{n \in \mathbb{N}} (Tn, \{m : m \; E \; n\}).$$

It is mechanical to verify that the set of finite V_n in this model is the set of all natural numbers of the form $\beth_n - 1$ where $\beth_n$ is the nth **finite** beth number. Thus $\beth_0 = 1$ and $\beth_{n+1} = 2^{\beth_n}$. Let this set be Y. We must show that $V^\pi \models (\Lambda \in Y) \wedge (\mathcal{P}``Y \subseteq Y)$.

Suppose $k \in Y$. We want Y to contain some integer which is $\mathcal{P}`k$ in the sense of V^π. The desired integer will in fact be $T(2^{Tk+1} - 1)$, for $z \in_\pi T(2^{Tk+1} - 1)$ iff the Tzth bit of $T(2^{Tk+1} - 1)$ is 1, which is to say that $z < k$. But in general, if x is one less than a power of 2, then $y \leq x$ is the same as $(\forall u)(uEy \to uEx)$, so in this case we infer $(\forall u)(uEz \to uEk)$. That is, $V^\pi \models z \subseteq k$. The inferences are all reversible, so we infer that $T(2^{Tk+1} - 1)$ is the power set of Tk in the sense of V^π.

What remains to be shown—under the assumption of $\text{AxCount}_\leq$ in the form that $<^T$ is well-founded—is that in V^π the set Y is $\subseteq$-minimal among the sets containing Λ and closed under $\mathcal{P}$.

Let y be a set such that $V^\pi \models \Lambda \in y \wedge \mathcal{P}``y \subseteq y$. y is clearly fixed. We want to show that $\beth_n - 1 \in y$.

Let $T(2^{k+1}) - 1$ be one-less-than-a-beth-number which is not in y and $<^T$-minimal with this property. But if $T(2^{k+1}) - 1$ is not in y, then k cannot be either, and $k <^T T(2^{k+1}) - 1$ contradicting $<^T$-minimality. ■

We can use this to elaborate on the conjectures about the class Γ of sentences such that $WF \prec_\Gamma V$ seen in section 2.1.4. If $\neg\text{AxCount}_\leq$, then by proposition 3.1.16 we have a permutation model containing a finite $\mathcal{P}`x \subseteq x$, so that the axiom of infinity is false in WF and $WF \not\prec_{str(\Sigma_1^{\mathcal{P}})} V$. Therefore

$$A: \quad \neg\text{AxCount}_\leq \to \Diamond(WF \not\prec_{str(\Sigma_1^{\mathcal{P}})} V).$$

Since no-one has yet discovered a permutation which we can use to get rid of infinite well-founded sets the possibility is open that $\text{AxCount}_\leq \to (\exists x \in WF)(x \text{ infinite})$ (since we can make the consequent true in some permutation model if the antecedent holds), or even

$$\text{AxCount}_\leq \to (WF \prec_{str(\Sigma_1^{\mathcal{P}})} V).$$

The most extreme conjecture in this direction would be the converse to A, so I offer as an interesting possibility to be confirmed or denied:

$$\text{AxCount}_\leq \longleftrightarrow \Box(WF \prec_{str(\Sigma_1^{\mathcal{P}})} V).$$

We have seen earlier that there are no definable infinite well-founded sets, so in particular there are none in any term model of *NF*. If it is indeed the case that there is no way of getting rid of them by permutations, then $\text{AxCount}_{\leq}$ would have to be false in all term models; otherwise, starting from a term model we could add a well-founded infinite set by permutations and then claim (since there is no way of getting rid of it) that it had been there all along. If $\text{AxCount}_{\leq}$ is false in all term models, this would mean that in every term model there is a closed term (a set abstract) denoting a non-standard integer. This seems unlikely, but is not obviously false.

3.1.1.3 *The action of J_0 on the family of permutation models* The most obvious task in the study of permutation models is to find ways of using information about J_0 (the internal symmetric group on V) to derive consistency and independence results in a more systematic way than has been possible so far. Unfortunately the general drift of the results proved in this section and the next is that group-theoretical information about individual permutations σ has nothing to tell us about what will be true in V^σ, nor anything about embeddings between V^σ and other V^τ.

Define an action of J_0 on itself by $\langle \sigma, \pi \rangle \mapsto (j`\sigma)\pi\sigma^{-1}$. Let us call this *skew-conjugation.* The significance of it is as follows.

THEOREM 3.1.29 *$V^\sigma \simeq V^\tau \longleftrightarrow \sigma$ and τ are skew-conjugate.*

Proof: (Here $\simeq$ means that the isomorphism is a *set.*) $x \in \tau`y$ iff $\sigma`x \in (j`\sigma)`\tau`y$ iff $\sigma`x \in (j`\sigma)\tau`\sigma^{-1}\sigma`y$ which is to say that $\sigma$ is an isomorphism between $V^\tau$ and $V^{(j`\sigma)\tau`\sigma^{-1}}$.

There is an exactly similar theorem for setlike permutations, saying that, if τ and σ are setlike permutations so that V^τ and V^σ are isomorphic (where the isomorphism is setlike), then τ and σ are skew-conjugated by a setlike permutation. ■

The following theorem was motivated by the observation that until people started using proposition 2.1.10 (which does not necessarily output involutions), all the permutations in proofs of consistency of formulae with respect to *NF* were in fact involutions. This raised the suspicion that one need not consider permutations of any other order. What we want is a theorem that tells us what constraints there can be on permutations that make Φ true if it was not true already, and we will get the desired result on involutions as a corollary.

THEOREM 3.1.30 *Let $A(\ )$ be stratified; define a normal generating subset of J_0 (i.e. $\{\sigma : A(\sigma)\}$ is a union of conjugacy classes generating J_0); and suppose that*

$$NF \vdash \neg\Phi \to (\forall\sigma)(\Phi^\sigma \to \neg A(\sigma)).$$

Then Φ *is invariant.*

Proof: Suppose we have proved '$\neg\Phi \rightarrow (\forall\sigma)(\Phi^\sigma \rightarrow \neg A(\sigma))$'. Then we have proved '$(\forall\tau)(\neg\Phi \rightarrow (\forall\sigma)(\Phi^\sigma \rightarrow \neg A(\sigma)))^\tau$', which we can simplify (since (σ is a permutation)$^\tau$ iff τ_n'σ is a permutation, for n sufficiently large we can take a shortcut by using this relettering and not spelling out quantified Greek letters): '$(x \in \sigma\text{`}y)^\tau$' is equivalent to '$\tau_{n-1}\text{`}x \in (\tau_{n+1}\text{`}\sigma)\tau_n\text{`}y$' for n sufficiently large. We can expand the right hand side to

$$\tau_{n-1}\text{`}x \in (\tau_n)(\tau_n^{-1})(\tau_{n+1}\text{`}\sigma)\tau_n\text{`}y$$

and use again the fact that, for all u and v, $u \in \tau\text{`}v$ iff $\tau_n\text{`}u \in \tau_{n+1}\text{`}v$, to get

$$x \in \tau(\tau_n^{-1})(\tau_{n+1}\text{`}\sigma)\tau_n\text{`}y.$$

Now $A(\ )$ is by hypothesis stratified, so '$A(\sigma)^\tau$' becomes '$A(\tau_{n+1}\text{`}\sigma)$' with $n$ sufficiently large. Now we can reletter '$(\tau_{n+1}\text{`}\sigma)$' as '$\sigma$', getting

$$x \in \tau(\tau_n^{-1})\sigma\tau_n\text{`}y \text{ and } A(\sigma).$$

Now we use the fact that A is normal, so that $A(\sigma) \longleftrightarrow A((\tau_n^{-1})\sigma\tau_n)$; thus, if we reletter $(\tau_n^{-1})\sigma\tau_n$ as σ and use substitutivity of the biconditional, we get $x \in \tau\sigma\text{`}y$ and $A(\sigma)$. This sequence of relettering lies at the core of all proofs in this section. We have thus turned the whole expression into

$$(\forall\tau)(\neg\Phi^\tau \rightarrow (\forall\sigma)(\Phi^{\tau\sigma} \rightarrow \neg A(\sigma)))$$

or, more readably,

$$(\forall\tau)(\forall\sigma)((\neg\Phi^\tau \wedge A(\sigma)) \rightarrow \neg\Phi^{\tau\sigma}).$$

That is to say, we have a proof that the collection of τ such that $\neg\Phi^\tau$ is closed under postmultiplication by things which are $A(\ )$. ("Collection" here is deliberately ambiguous between "set" and "class": such collections may or may not be sets, and in any case it does not matter here.) So if $\exists\tau\neg\Phi^\tau$ then, for all σ, since A is a set of generators, $\tau^{-1}\sigma$ is a product of things in A, so $\neg\Phi^{\tau(\tau^{-1}\sigma)}$ as well, $\forall\sigma\neg\Phi^\sigma$, which is to say ϕ is invariant. ■

This has a multiplicity of corollaries. For example, there is no closed wff ϕ such that $(\sigma^2 = 1) \longleftrightarrow \phi^\sigma$.

There is no reason to suppose that theorem 3.1.30 is the best possible. It might well be the case that every skew-conjugacy class meets every conjugacy class. It is also open whether or not ϕ^σ can be equivalent to a stratified property of σ without ϕ being invariant.

As we saw on page 102 some fairly weak AC (to wit, the axiom GC) is enough to imply that involutions comprise a normal generating subset, so we may take $A(\sigma)$ to be $\sigma^2 = 1$, and conclude as desired that we can never be proved to need a permutation of order other than two.

3.1.1.4 *Relations between permutation models from a common base set*

THEOREM 3.1.31 *In NF all permutation models from a common base set have the same definable sets.*

Proof: This needs clarification. For present purposes, the most useful notion of definable is *symmetric*. Suppose x and y are two symmetric sets in a model $\langle M, \in\rangle$, and let τ be an (internal) permutation of M. For $k_1, k_2 > n$, where x is n-symmetric, we have

$$\tau(j`\tau)(j^2`\tau)(j^3`\tau)\ldots(j^{k_1}`\tau)`x = \tau(j`\tau)(j^2`\tau)(j^3`\tau)\ldots(j^{k_2}`\tau)`x$$

since x, being n-symmetric, is fixed by $j^k`\tau$ for $k \geq n$. Let us write this eventually constant value as "$\sigma`x$", bearing in mind that σ is probably not going to be a set. Now, for any x and y, we have $x \in y$ iff

$$\tau(j`\tau)(j^2`\tau)(j^3`\tau)\ldots(j^n`\tau)`x \in (j`\tau)(j^2`\tau)(j^3`\tau)\ldots(j^{n+1}`\tau)`y$$

which is the same as

$$\tau(j`\tau)(j^2`\tau)(j^3`\tau)\ldots(j^n`\tau)`x \in_{\tau^{-1}} \tau(j`\tau)(j^2`\tau)(j^3`\tau)\ldots(j^{n+1}`\tau)`y.$$

So if y is symmetric, and n is large enough,

$$\tau(j`\tau)(j^2`\tau)(j^3`\tau)\ldots(j^{n+1}`\tau)`y = \tau(j`\tau)(j^2`\tau)(j^3`\tau)\ldots(j^n`\tau)`y.$$

That is to say, $x \in y$ iff $\sigma`x \in_{\tau^{-1}} \sigma`y$. We now show that $\sigma`x$ is symmetric in the sense of $M^{\tau^{-1}}$ iff $x$ was symmetric in $M$, and this will show that $\sigma$ is not only an isomorphism between $M$ and $M^{\tau^{-1}}$ but sends symmetric sets to symmetric sets. Now "$x$ is $n$-symmetric" is a stratified expression,[47] so "$\sigma`x$ is n-symmetric$^{\tau^{-1}}$" is simply "$((\tau^{-1})_k)\sigma`x$ is $n$-symmetric". But $((\tau^{-1})_k)\sigma`x$ is simply x. ■

At the beginning of this chapter, I promised to look at embedding relations between permutation models. These are relations between τ and σ which assert

$$(\exists f : V \to V)(f \text{ is 1-1} \wedge (\forall \vec{x})(\Phi(\vec{x})^\sigma \longleftrightarrow \Phi(f`\vec{x})^\tau))$$

where Φ is some suitable predicate. As long as the first-order theory of an embedding relation is invariant we can pretend that we are looking at J_0

[47] With *one* free variable only: 'n' is not a variable!

from outside. It would be grossly pathological if the individual permutation models made differing allegations about the family of all permutation models, for then we could see from outside that some were right and others wrong! Fortunately this does not arise.

PROPOSITION 3.1.32 *The first-order theory of any embedding relation is invariant.*

The proof in Forster [1987a] has not been improved on so far, and there is little point in recapitulating it here. I stated and proved it only for first-order theories of embedding relations: in fact the same proof will work for the nth-order theory, for each n. Although we now know that the theory of any embedding relation is invariant, we still have absolutely no idea what any of them contain! The next result makes this explicit.

PROPOSITION 3.1.33 *Consider a language extending set theory with new primitives for various embedding relations as above. Let* $\Phi(\ )$ *be some expression with one free (Greek) variable in this language, and let* $A(\ )$ *be normal and stratified.*

Let us suppose

$$\vdash (\exists\sigma)(\Phi(\sigma))$$

and

$$\vdash (\forall\sigma)(\Phi(\sigma) \to A(\sigma)).$$

Then

$$\vdash (\forall\sigma)(A(\sigma)).$$

In other words, facts about embedding relations between permutation models tell us nothing about the group-theoretic properties of the permutations involved.

Proof: We have $\vdash (\forall\sigma)(\Phi(\sigma) \to A(\sigma))$ so we will have $\vdash (\forall\tau)(\forall\sigma)(\Phi(\sigma) \to A(\sigma))^{\tau}$. With the customary relettering for σ, this will become $\vdash (\forall\tau)(\forall\sigma)(\Phi(\tau\sigma) \to A(\sigma))$. So as long as something is Φ, everything must be A. ∎

We find ourselves in a situation very similar to that in which theorem 3.1.30 placed us: we would like to have a stronger version (with Φ, A, as above):

$$[(\exists\sigma)(\Phi(\sigma)) \wedge (\forall\sigma)(\Phi(\sigma) \to A(\sigma))] \to (\forall\sigma)(A(\sigma))$$

but, as before, the technique of the proof which gave us the weak version will not deliver the strong. This time I will omit the proof that the strong version is at least invariant.

3.1.2 *Outer automorphisms in NF*

Hitherto in this chapter we have restricted ourselves to setlike permutations which are *sets*. Although *NF* is a set theory with a universal set, so that permutations of a model of *NF* may be sets of the model (so this is not as expensive a restriction as it would be with *ZF*!), Henson's lemma requires only that the permutations involved should be setlike. Outer $\in$-automorphisms of M are certainly setlike, so we can use them too: they will be the subject of this section. Since outer automorphisms are not sets, we cannot modalize the subject matter the way we could when we were considering permutations that were sets: this is not iterable. However, just as there were advantages in restricting attention to permutations that are sets (the smooth-running modal theory) there are advantages too in restricting ourselves to outer automorphisms: the Ehrenfeucht–Mostowski theorem is a rich source of models with many outer automorphisms. All applications of this are due to Henson [1969, 1973b], Körner [1994], and Pétry [1975, 1976, 1977, 1979], and the reader is urged to consult these references.

One advantage of this approach is that, when τ is an outer automorphism, then $j`\tau = \tau$, so $\tau_n = \tau^n$ for every n. Now τ^n is a much easier object to visualize than τ_n, so this approach is easier to follow. Also, since these are *outer* automorphisms, the construction preserves *stratified* sentences only, not all invariant ones as well.

Since, by the results of the previous section, cardinal arithmetic is invariant, if we wish to obtain independence results in it by permutations we must use outer setlike permutations. In fact the first use of this construction (theorem 2.1 of Henson [1969]) is to prove the independence of the axiom of counting, $\text{AxCount}_{\leq}$ and $\text{AxCount}_{\geq}$.

THEOREM 3.1.34 Henson [1969]. *Let S be any consistent stratified extension of NF, and $\langle B, \prec\rangle$ any linear order. Let $B = B_1 \sqcup B_2 \sqcup B_3$ be any partition of B. Then S has a model M such that $\langle B, \prec\rangle$ is embedded in the natural numbers of M*

and

if $x \in$ (image of B_1) then $M \models x < Tx$;
if $x \in$ (image of B_2) then $M \models x = Tx$;
if $x \in$ (image of B_3) then $M \models x > Tx$.

Proof: By the Ehrenfeucht–Mostowski theorem, we can form a model M so that $B \times \mathrm{Z}$ is embedded in the cantorian natural numbers, and such that every automorphism of $B \times \mathrm{Z}$ extends to an automorphism of M. Now consider the automorphism σ of $B \times \mathrm{Z}$ given by $\sigma`\langle b, n\rangle = \langle b, n-1\rangle$ for $b \in B_1$, $= \langle b, n\rangle$ for $b \in B_2$, $= \langle b, n+1\rangle$ for $b \in B_3$. Readers who have survived the chapter this far will have no difficulty verifying that, in M^σ, naturals n that started in $B_1 \times \mathrm{Z}$ satisfy $Tn < n$, naturals n that started in $B_3 \times \mathrm{Z}$ satisfy $Tn > n$, and naturals n that started in $B_3 \times \mathrm{Z}$ satisfy $Tn = n$. ∎

THEOREM 3.1.35 Pétry [1975]. *To any consistent stratified extension S of NF we can consistently add the assertion:*

$$\exists \alpha \in NC\ \alpha \not\geq^* T\alpha \not\geq^* \alpha \not\geq^* 2^{T\alpha} \not\geq^* \alpha.$$

Proof: We first prove that, for any $n \in \mathbb{N}$, we can find a sequence of cardinals $\langle \alpha_i : i < n \rangle$ such that, for $0 \leq i < j \leq n$, $T\alpha_j \not\leq^* \alpha_i \not\leq 2^{T\alpha_j}$. Theorem 2.2.4 states that $\overline{\overline{NO}} \leq T\ \overline{\overline{V}}$ but $\overline{\overline{NO}} \not\leq T^2\ \overline{\overline{V}}$. Pétry provides the following delicate and ingenious construction: for $i \leq n$, set

$$\beta_i = T^{4n+1-2i}\ \overline{\overline{V}},\quad \gamma_i = T^{2i-1}\ \overline{\overline{NO}},\ \text{and}\ \alpha_i = \beta_i + \gamma_i.$$

The α_i are in fact what we want, and this is what we must prove next. Suppose $0 \leq i < j \leq n$. Then:

1. Obviously (count the exponent) $T\beta_j \not\leq^* \beta_i$.
2. $T\beta_j \not\leq^* \gamma_i$; otherwise V would be well-ordered.
3. $T\alpha_j \not\leq^* \alpha_i$; otherwise $(T\beta_j)^2 = T\beta_j \leq^* \beta_i + \gamma_i$, which contradicts 1 and 2 if we use Bernstein's lemma $(m + n \geq^* p.q \to m \geq^* p \vee n \geq^* q)$.
4. $\alpha_j \leq T^{2j}\ \overline{\overline{V}}$ (count the exponent).
5. $\alpha_i \not\leq^* 2^{T\alpha_j}$; if not, then $\gamma_i \leq^* T^{2j}\ \overline{\overline{V}}$ (by 4) and $2^{T\gamma_i} \leq T^{2j}\ \overline{\overline{V}}$, so $T\gamma_i \leq T^{2j}\ \overline{\overline{V}}$, so $\overline{\overline{NO}} \leq T^2\ \overline{\overline{V}}$, which we know to be false.

By compactness we can blow this up to a Z-string of such cardinals and invoke the Ehrenfeucht–Mostowski theorem. ∎

PROPOSITION 3.1.36 Pétry [1979] *Let S be a consistent stratified extension of NF. Then we can consistently add to S two constants a and b, with axioms $\overline{\overline{a}} < \overline{\overline{\mathcal{P}`a}}$, $\overline{\overline{\mathcal{P}`b}} < \overline{\overline{b}}$, and axiom schemes $\Phi(a) \longleftrightarrow \Phi(V)$ and $\Phi(b) \longleftrightarrow \Phi(V)$ for all typed Φ.*

Proof: If Φ is a stratified formula, let Φ_x be the typed formula obtained by restricting to $\mathcal{P}^{n}`x$ all variables in Φ receiving the type n in the canonical stratification. Let L be the theory of linear order all of whose axioms are relativized to a one-place predicate $\mathcal{U}$ and let $\mathcal{R}(x, y)$ be the binary relation symbol of L. Let S' be the union of $S \cup L$ and the following:

(i) $\forall x(\mathcal{U}(x) \to \overline{\overline{x}} = \overline{\overline{\mathcal{P}`x}})$

(ii) $\forall x((\forall y)(\mathcal{R}(x, y) \to \overline{\overline{x}} \leq \overline{\overline{\iota``y}}\ ))$

(iii) $\forall x(\mathcal{U}(x) \to (\Phi \longleftrightarrow \Phi_x))$ for all stratified Φ.

Let $\langle M, \in_M \rangle \models S$. Let $U = \{x \in M : (\exists k \in \mathbb{N})(\langle M, \in_M \rangle \models x = \iota^{2k}``V)\}$ and let $x\ R\ y$ iff

$$(\exists k < l \in \mathbb{N})(\langle M, \in_M \rangle \models x = \iota^{2l}``V \wedge \langle M, \in_M \rangle \models y = \iota^{2k}``V).$$

Since all $\iota^n``V$ are the same size as their power sets, the structure $\langle M, \in_M, U, R \rangle$ satisfies (iii). Also $\langle U, R \rangle$ is an infinite linear order, so $\langle M, \in_M, U, R \rangle \models S'$. Now we use the Ehrenfeucht–Mostowski theorem to obtain a model $\langle N, \in_N, U', R' \rangle$ of S' with an automorphism σ and two elements a and b so that $\sigma`a \; R \; a$ and $b \; R \; \sigma`b$, and $\langle N, \in_\sigma \rangle$ is the model we need. ■

We conclude this section with a statement (but not a proof) of the most recent result obtained by these methods.

THEOREM 3.1.37 Körner [1994]. *If NF is consistent it has a model in which for all sufficiently large $n \in \mathbb{N}$, $n < Tn$.*

3.2 Applications to other theories

Since this is a work on set theory with a universal set, it is not the place for an exhaustive treatment of applications of this method to *ZF*. Apart from replacement, the axioms of *ZF* are all stratified and so are preserved under permutation models. In fact the axiom scheme of replacement is also preserved as long as the permutation is definable (see Rieger [1957]). It is fairly simple to prove the following.

REMARK 3.2.1 *If M is a well-founded model of ZF and σ a setlike permutation of M, then M and M^σ contain the same well-founded sets.*

Proof: Define i recursively by: $i`x = \sigma^{-1}`(i``x)$. Clearly i sends M into M^σ. Since everything in M has only an M-set of $\in_\sigma$-members, we can keep the recursion going so that every subset (in the sense of $\in_\sigma$) of the range of i is also in the range of i. But if M^σ contained new well-founded sets, there would have to be minimal ones not in the range of i, and we have just seen that this is not possible. ■

This last result explains why Rieger–Bernays permutation models are no use in *ZF*: the only way of getting a model that is not isomorphic to the one you started with is to add some sets that are not well-founded. If we choose to, by varying the construction slightly, we can even add a universal set. This construction is the subject of the next chapter.

4

CHURCH–OSWALD MODELS

In [1974] Church showed us a consistency proof for an axiomatic system of set theory containing *inter alia* an axiom to the effect that there is a set of all sets. This bore no relation to any previously published set theories with such an axiom, and naturally involved a new construction. For people who wish to inform themselves about this construction and the set theories whose consistency it proves, Church's original paper is not ideal. Its purpose was to communicate an *aperçu* rather than give a comprehensive treatment. The proofs are not provided in any detail (Church says the proof is "simple but ... laborious"!), and the construction he uses is an extremely complicated instance of the method he introduces. At about the same time Urs Oswald in Zürich independently discovered the same construction, and gave a much simpler use of it, which will be our starting point here, for the two come together very well.

Church-Oswald constructions are essentially interpretations of axiomatic set theories with a universal set inside (typically) ZF. Of course the axiom of foundation fails in the models we construct, but we will assume it holds in the set theory in which we work.

4.1 Oswald's model

We saw on page 113 a well-known trick (due to Ackermann) of defining a binary relation E on $\mathbb{N}$ so that $\langle \mathbb{N}, E\rangle \simeq \langle V_\omega, \in\rangle$. We say $n\ E\ m$ iff the nth bit of the binary expansion of m is 1. Oswald [1976] discovered that an easy twist to this—write instead nEm iff m is odd and the nth bit of the binary expansion of $(m-1)/2$ is 1 or m is even and the nth bit of the binary expansion of $m/2$ is 0—gives us a model of a set theory with a universal set, since evidently every "set" of the model now has a "complement".

Oswald's interest in this structure arose from the fact that it is a particularly pleasing and unexpected presentation of the term model of a theory called NF_2 which he was studying at the time.

How do we recover terms from numbers? By recursion. If m is odd it corresponds to the unordered n-tuple formed from the terms corresponding to the addresses in $(m-1)/2$ of non-zero bits. If m is even it corresponds to the complement of the term corresponding to $m+1$. So every number x corresponds to either an unordered tuple or the complement of an unordered tuple of sets corresponding to numbers less than x. Clearly this process is going to terminate.

Of course we can do this in a more general context. Take an arbitrary model of *ZF* (for the moment we do not need either foundation or choice) and consider a bijection between V and $V \times \{0,1\}$. Let us write this 'k' for '$k$ode', and define

$x \in_{new} y$ iff

1. $k`y = \langle y', 0\rangle$ and $x \in y'$ *or*
2. $k`y = \langle y', 1\rangle$ and $x \notin y'$.

This model, too, is a model of NF_2, but this time the well-founded part of it is at least as big as the original model, just as in Oswald's model there are still all the hereditarily finite sets. (Proofs of all of this will follow.)

Once we have the trick expressed in this form, we can see the proper way to generalize it. We took k to be a bijection between V and $V \times \{0,1\}$, but it could have been $V\times$ anything K we like ('K' for 'Kode'), as long as $0 \in K$ and we have suitable clauses for $\in_{new}$ x where the second component of $k`x$ is not 0. The idea is that in general we define $x \in_{new} y$ if either (i) $\texttt{snd}(k`y) = 0$ and $x \in \texttt{fst}(y)$ or[48] (ii) various other clauses concerning (new) membership in sets y such that $\texttt{snd}(k`y) \neq 0$. The idea is that other entries in K will correspond to other operations on sets (in the case we have just seen 1 corresponds to complementation). In general there is a problem of extensionality. There is of course no difficulty in showing that x and x' satisfy extensionality as long as $\texttt{snd}(k`x) = \texttt{snd}(k`x') = 0$ but there are other cases to consider. If the theory for which we are trying to obtain a model by this construction is T, then the extensionality problem for the model is deeply related to the word problem for T. In particular if we have a good notion of normal forms for T words over a set of generators then we will be able to take K to be (roughly) a set of such normal words. *What this reveals is that this technique is not a great deal of use for constructing models of a theory T unless T has an easy word problem.*

From now on until further notice we will pointedly ignore the question of what the other second components of $k`x$ can be. We will assume that $K$ is some arbitrary collection such that $0 \in K$ and there are rules to ensure that $\in_{new}$ is extensional and that when $\texttt{snd}`k`y = 0$ then $x \in_{new} y \longleftrightarrow x \in \texttt{fst}(k`y)$. We will try to prove some general results about this construction. Let us call structures built in this way **CO structures**. ("Church–Oswald").

To summarize, there are three parts to a CO construction over a base model $\langle V, \in\rangle$. There is a collection K of objects available to be used as second components of ordered pairs; there is a bijection k between V and $V \times K$; and finally there is a family of rules telling us how possible membership (in the new sense) of x in y depends on the second component of $k`y$.

[48] $\texttt{fst}(x)$ and $\texttt{snd}(x)$ are the first and second components of the ordered pair x.

(Having announced this as the general form of CO constructions we will then of course fail to respect it. In a number of cases the bijection k is not between V and $V \times K$ but between V and some subset of $V \times K$. This is the case in Church's model below, and the result of section 4.4.2. The flavour remains the same.)

4.2 Low sets

A **low set** is a set x such that $\mathtt{snd}(k`x) = 0$. This is at variance with other definitions in the literature, but it is the most useful. This is not to be confused with the notion of a **hereditarily low** set, for this will turn out to be important as well.

DEFINITION 4.2.1 *The axiom scheme of* **low comprehension** *states that, for any formula $\phi(x, \vec{y})$, and for all $\vec{y}$, if the collection of all x such that $\langle V, \in_{\mathbf{CO}}\rangle \models \phi(x, \vec{y})$ is a set of the original model, then $\langle V, \in_{\mathbf{CO}}\rangle \models$ "$\{x : \phi(x, \vec{y})\}$ is a set".*

Thus it is by no means obvious that low comprehension is axiomatizable. For this reason the most profitable approach to this topic is to think of the CO constructions as things that give us models, rather than to attempt to be specific in saying what the axioms are of the theory whose consistency we have proved.

The following triviality is central to what is to come.

THEOREM 4.2.2 *All CO structures satisfy low comprehension.*

Proof: Let $\phi(x, \vec{y})$ satisfy the antecedent, and consider the class of all x such that $\langle V, \in_{\mathbf{CO}}\rangle \models \phi(x, \vec{y})$ which is a set of the original model, X, say. Then $\{x : \phi(x, \vec{y})\}$ in the sense of the new model is simply $k^{(-1)}`\langle X, 0\rangle$. ■

In a typical CO construction there will be plenty of new sets containing all low sets: V for one. However,

PROPOSITION 4.2.3 *No new set containing all low sets can be low.*

Proof: Suppose there were a low set containing all low sets. Then, by low comprehension, the collection of all low sets is low. That is to say, there is an x such that $(\forall y)(y \in_{\mathbf{CO}} x \longleftrightarrow \mathtt{snd}(k`y) = 0)$. But there is certainly a proper class of $y$ such that $\mathtt{snd}`(k`y) = 0$, so this x has a proper class of members and is therefore not low. ■

But we can arrange for the collection of low sets to be a non-low set of the model. This is simple enough. Vary Oswald's construction very slightly. When m is odd we set $n \in_{\mathbf{CO}} m$ iff the nth bit of $(m-1)/2$ is 1 (so that odd numbers code finite sets). When m is even and at least 4 we set $n \in_{\mathbf{CO}} m$ iff the nth bit of $(m-4)/2$ is 0. This uses large even numbers to code cofinite sets and leaves 0 and 2 free to code two extra things, so we set $n \in_{\mathbf{CO}} 0$ iff

n is odd, and $n \in_{\mathbf{co}} 2$ iff n is even. In the new model 0 is now the set of all finite (i.e. low) sets.

COROLLARY 4.2.4 *Every surjective image of a low set is low and every subset of a low set is low.*

Proof: The first part follows from replacement in the original model and the second from low comprehension. ■

COROLLARY 4.2.5 *Every low set has a power set, which is also low.*

Proof: When x is low, $\mathcal{P}`x$ (in the sense of $\in_{\mathbf{co}}$) must be

$$k^{(-1)}`\langle k^{(-1)}``(\mathcal{P}`(\mathtt{fst}(k`x)) \times \iota`0), 0\rangle.$$

■

4.2.1 *Other definitions of low*

Another natural definition of "low" is: a set x is low if the collection $X = \{y : y \in_{\mathbf{co}} x\}$ is a set of the old model (x has "only a set" of members in the sense of $\in_{\mathbf{co}}$). It is an immediate consequence of extensionality that this definition is equivalent to our original definition. Let x and X be as in the new definition. Then $k^{(-1)}`\langle X, 0\rangle$ and x have the same members-in-the-sense-of-$\in_{\mathbf{co}}$, and so, by extensionality-for-$\in_{\mathbf{co}}$, must be the same object.

Church defines a low set to be something the same size as a well-founded set, and Sheridan follows him in this. This is not a useful definition, for it does not enable us to get the slick proof of theorem 4.2.2 we have just seen. What is a well-founded set anyway? The trouble with definition 2.1.17 as a definition of well-founded sets is that if we have very little comprehension the only y such that $\mathcal{P}`y \subseteq y$ may be the new V in which case *everything* is well-founded. Saying a set is well-founded if there is an $\in$-homomorphism from $TC(x)$ onto some von Neumann ordinal is no use, because without adequate comprehension we cannot be sure that either $TC(x)$ or the requisite von Neumann ordinals exist. If we really wish to discuss well-founded sets of the new model, there seems to be no alternative to a definition that involves bound class variables, such as a version of definition 2.1.17 where the intersection is taken over proper classes. Because of the separation scheme in ZF, we do not need bound class variables to discuss well-foundedness, and it is not clear how gross a pathology their appearance here should be held to be.

4.3 $\mathcal{P}$-extensions and permutation models

4.3.1 $\mathcal{P}$-extensions

As long as the original model $\langle V, \in \rangle$ was well-founded there is a canonical injection from it into the new model $\langle V, \in_{\mathbf{co}} \rangle$ defined by recursion on $\in$:

DEFINITION 4.3.1 $i`x =_{\text{df}} k^{(-1)}`\langle i``x, 0 \rangle$.

THEOREM 4.3.2 *If $\langle V, \in \rangle$ is well-founded then i is defined and is a $\mathcal{P}$-embedding from $\langle V, \in \rangle$ into $\langle V, \in_{\mathbf{co}} \rangle$.*

Proof: We must first check that i is an isomorphism so we want $i`x \in_{\mathbf{co}} i`y \longleftrightarrow x \in y$. Since the second component of $k`(i`y)$ is 0, $i`x \in_{\mathbf{co}} i`y$ iff $i`x \in \mathtt{fst}(k`(i`y)) = i``y$ iff $x \in y$.

Next we show that the range of i is transitive$^{(\in_{\mathbf{co}})}$. Suppose y is in the range of i, and $y = i`z$. So $x \in_{\mathbf{co}} i`z = k^{(-1)}`\langle i``z, 0\rangle$ iff $x \in i``z$ so x would also be in the range of i.

Finally we must check that any subset of something in the range of i is likewise in the range of i. Suppose x is in the range of i, so that $k`x = \langle i``z, 0\rangle$ for some $z$. Suppose also that $(\forall w)(w \in_{\mathbf{co}} y \to w \in_{\mathbf{co}} x)$. Consider the set (in the sense of the original model) of those things that are $\in_{\mathbf{co}}$ $y$. This is indeed a set of the original model, since it is a subset of $i``z$. If it is $i``u$, then $k`y$ must be $\langle i``u, 0\rangle$ so y is in the range of i. ■

(Essentially this proof is in Church [1974] though he prefers to allege that the well-founded sets form a model of *ZF*, and does not have the concept of a $\mathcal{P}$-embedding.)

Notice that it is not being claimed that in general there are no new well-founded sets, nor that the range of i is not a set of the new model. This will be discussed more in section 4.3.3. It does not seem possible to obtain a contradiction uniformly from the assumption that the range of i is coded in the new model by some $\langle y, x\rangle$ where x is not 0, though in some cases we can.

PROPOSITION 4.3.3 *In Oswald's model the range of i is not a set.*

Proof: We want to show that if X satisfies $(\forall y \in_{\mathbf{co}} X)$($y$ is in the range of i) then X is in the range of i too. The case $\mathtt{snd}(k`X) = 0$ we have already considered. There remains the case $\mathtt{snd}(k`X) = 1$.

We will show that this case cannot occur. If it did, $x \in_{\mathbf{co}} X$ iff $x \notin \mathtt{fst}(k`X)$. So $X$ would have cofinitely many members$^{(\in_{\mathbf{co}})}$, all of them values of $i$. But there are not cofinitely many values of $i$ since if $x \in$ range $i$, $\mathtt{snd}(k`x) = 0$, and there are infinitely many x such that $\mathtt{snd}(k`x) = 1$. Therefore $\mathtt{snd}(k`X) \neq 1$. ■

PROPOSITION 4.3.4 *We can also arrange that the range of the canonical embedding* is *a set of the new model (though of course not a low one).*

Proof: We do this by a simple modification of Oswald's original model rather like that we saw just before the beginning of section 4.2.1. Since for something to be in the range of i it is sufficient for its last bit to be hereditarily 0, this means that we can—given k—recognize integers that will eventually be in the range of i without ever knowing what the membership conditions for odd numbers are. This in turn means that we can then organize the membership conditions so that even numbers code finite sets, 1 codes the set of all things in the range of i, 3 codes its complement, and the remaining odd numbers code cofinite sets as before. ■

Why isn't it obvious that there are no new well-founded sets? After all, they would have to be of rank bigger than anything we had in the old model, and then surely there would be some new inaccessible cardinal and a consistency proof violating the second incompleteness theorem. But life is not that simple: to get a consistency proof we would need enough comprehension (e.g. $\in$ restricted to old sets to be a set) and this we just do not have *in general.*

4.3.2 *Hereditarily low sets and permutation models*

Hlow is the class of hereditarily low sets[49]—in the sense of $\in_{\mathbf{co}}$ of course. One might expect that this should turn out to be identical with (i) the range of the canonical embedding i, or (ii) with the class of sets that are well-founded in the sense of $\in_{\mathbf{co}}$. Leaving aside the obscurities in the notion of well-foundedness-in-the-sense-of-$\in_{\mathbf{co}}$ (see section 4.2.1) we can see almost immediately that since we have a lot of freedom in deciding how to *k*ode pairs as sets it might happen that the new model contains Quine atoms. Such objects are clearly hereditarily low and are not well-founded, so (ii) looks extremely unlikely. (We shall see later that it is actually falsifiable.) In these circumstances (i) fails too: i is an $\in$-isomorphism defined by recursion on $\in$ so nothing in its range can be self-membered. However, something like (i) and (ii) must be true, and the Quine atoms counterexample should enable *NF*istes to predict the statement of the next theorem.

THEOREM 4.3.5 (*AC*)

1. *Hlow is always isomorphic to a permutation model of the original universe.*
2. *Whatever K we started with, for any permutation σ of the old universe we can find a coding function k so that $\langle Hlow, \in_{\mathbf{co}}\rangle \simeq \langle V, \in_\sigma\rangle$.*

Proof:

[49] I have written '*Hlow*' rather than H_{low} because we do not want the least fixed point but the greatest.

(1) There will be a bijection $\pi : V \longleftrightarrow Hlow$. This follows from AC since $Hlow$ is a proper class. We seek a σ so that

$$(\forall xy)(x \in \sigma`y. \longleftrightarrow .\pi`x \in_{\mathbf{co}} \pi`y).$$

What is $\sigma`y$? Clearly it has to be $\{x : \pi`x \in_{\mathbf{co}} \pi`y\}$. This is a set, since $\pi`y$ is low. We must check that this definition gives us a σ that is 1-1 and onto. It is certainly 1-1 by extensionality of $\in_{\mathbf{co}}$. Is it onto? Given z we must find a y so that $z = \{x : \pi`x \in_{\mathbf{co}} \pi`y\}$. This y must be $\pi^{-1}`k^{-1}`\langle\pi``z, 0\rangle$.

(2) We know—however we choose k—that $Hlow$ is a proper class whose complement is a proper class, so let π be a bijection between V and such a class and let us fasten on that class to be $Hlow$ and resolve to cook up k so that it actually *is* the $Hlow$ of the new model. Dugald Macpherson has used the word "moiety" for things that are both infinite and coinfinite: we will borrow it here to describe proper classes whose complements are proper classes. Let σ be a permutation of V. We want to cook up k so that $\langle Hlow, \in_{\mathbf{co}}\rangle \simeq \langle V, \in_\sigma\rangle$. We want

$$(\forall xy)(x \in \sigma`y. \longleftrightarrow .\pi`x \in_{\mathbf{co}} \pi`y).$$

The right hand side is $\pi`x \in \mathtt{fst}(k`(\pi`y))$ (and $\mathtt{snd}(k`\pi`y) = 0$ since $\pi`y$ is a low set). Now $\mathtt{fst}(k`\pi`y) = \pi``\sigma`y$ so we want $k`\pi`y = \langle\pi``\sigma`y, 0\rangle$. It is true that this only tells us what k should do to values of π but since the range of π is a moiety and the range of $k \circ \pi$ is also a moiety there will be no problem extending this to a bijection between V and all the ordered pairs we need. ■

COROLLARY 4.3.6 *$\langle Hlow, \in_{\mathbf{co}}\rangle$ and $\langle V, \in\rangle$ are elementarily equivalent w.r.t. stratified formulae.*

There remains a rather obvious conjecture that I have not been able to prove. Is the $\mathcal{P}$-embedding $i : \langle V, \in\rangle \hookrightarrow^{\mathcal{P}}_{e} \langle Hlow, \in_{\mathbf{co}}\rangle$ elementary for stratified formulae? This is related to a question of Dzierzgowski's recently solved affirmatively by Friederike Körner [1994]: is it possible to have two models $\mathcal{A}$ and $\mathcal{B}$ of *NF* with $\mathcal{A}$ a substructure of $\mathcal{B}$ elementary for stratified formulae but $\mathcal{A}$ and $\mathcal{B}$ are not elementarily equivalent?

Some things we do know about this, and they constitute proposition 4.3.7. If we want a notion of well-founded that is geared to proving things about well-founded sets by $\in$-induction, then a set is to be deemed well-founded iff it belongs to all **classes** y such that every subset of y is a member of y.

PROPOSITION 4.3.7 *Consider the following properties:*[50]

[50]In the sense of $\in_{\mathbf{co}}$ of course.

1. *well-founded*
2. *hereditarily low*
3. *in the range of i.*

Neither of 1 and 2 imply the other, but their conjunction is equivalent to 3.

Proof:

$1 \not\rightarrow 2$ follows from proposition 4.3.4. This is because if the collection of things in the range of i is a set it must be well-founded, and it is not low (as we have seen) so *a fortiori* it cannot be in $Hlow$ either.

$2 \not\rightarrow 1$. By theorem 4.3.5 we can arrange for there to be Quine atoms in $Hlow$ and there cannot be any in WF.

To show $1 \wedge 2 \rightarrow 3$ we prove $2 \rightarrow 3$ by $\in$-induction. Consider the property of x: "if x is hereditarily low then x is in the range of i". Suppose this holds for all members of y and that y is hereditarily low. Then y is a low set of values of i and is therefore itself a value of i. Therefore the $\in$-induction works and we have a proof of $(\forall x)(WF(x) \rightarrow$ if x is hereditarily low then x is in the range of i).

$3 \rightarrow 1$ and $3 \rightarrow 2$ are easy. ■

4.3.3 *Permutation models of CO structures*

THEOREM 4.3.8 *For a given choice of K and V and rules, all CO structures are permutation models of each other.*

Proof: Suppose we have two CO structures over a model $\langle V, \in \rangle$, with the same K but different coding functions k and k' respectively. We wish to find $\sigma \in Symm(V)$ such that the first model thinks that $x \in y$ iff the second thinks $x \in \sigma`y$. σ must be $(k')^{-1} \circ k$. ■

We also have the following:

PROPOSITION 4.3.9 *Every permutation model $\langle V, \in_{\mathbf{co}} \rangle^\sigma$ of $\langle V, \in_{\mathbf{co}} \rangle$ is obtained from it by replacing k by some k', with a corresponding new membership relation $\in_{\mathbf{co}'}$. If the permutation is σ, then the new k' is $\sigma^{-1}k$.*

Proof:

$$\langle V, \in_{\mathbf{co}} \rangle^\sigma \models x \in y$$

iff

$$\langle V, \in_{\mathbf{co}} \rangle \models x \in \sigma`y$$

$$\langle V, \in \rangle \models x \in_{\mathbf{co}} \sigma`y$$

$$\langle V, \in \rangle \models x \in_{\mathbf{co}'} y$$

$$\langle V, \in_{\mathbf{co}'} \rangle \models x \in y.$$

■

4.4 Two applications

The burden of the last section was roughly this: if T is an axiomatic theory of sets that admits a universal set and has a very simple word problem, then any well-founded model of *ZF* has a $\mathcal{P}$-extension that is a model of T (and we can probably stipulate that the new model has no new well-founded sets, though this has not been specifically investigated). Now we are going to look at some particular constructions other than the Oswald toy we started with. The real problem is always extensionality, and although the reader has been warned that this problem is roughly equivalent to the word problem for the theory involved, it is perhaps helpful to work through a couple of these proofs to see just how grubby they can be.

4.4.1 *An elementary example*

We are going to consider two models, more complicated than the original Oswald model, in which in addition to complements for all x, we also have $B`x$ for all x. The second model will satisfy all the boolean axioms (and so is a model of *NFO*), but we start with the first, which doesn't. Consider $\mathcal{S} =$ the set of reduced words in the semigroup-with-unity with two generators $\mathtt{c}$ and $\mathtt{b}$, and the equation $\mathtt{c}^2 = 1_{\mathcal{S}}$. '$\mathtt{c}$' is intended to recall "Complement" and '$\mathtt{b}$' to recall "$B`x$". We now let k be a bijection between V and $V \times \mathcal{S}$. We define $\in_{\mathbf{co}}$ by recursion by cases:

DEFINITION 4.4.1

1. *If* $\mathtt{snd}(k`x) = \mathtt{c}w$ *for some word* $w \in \mathcal{S}$ *then* $y \in_{\mathbf{co}} x$ *iff* $y \notin_{\mathbf{co}} k^{(-1)}`\langle \mathtt{fst}(k`x), w\rangle$.
2. *If* $\mathtt{snd}(k`x) = \mathtt{b}w$ *for some word* $w \in \mathcal{S}$ *then* $y \in_{\mathbf{co}} x$ *iff* $k^{(-1)}`\langle \mathtt{fst}(k`x), w\rangle \in_{\mathbf{co}} y$.
3. *If* $\mathtt{snd}(k`x)$ *is* $1_{\mathcal{S}}$ *then* $y \in_{\mathbf{co}} x$ *iff* $y \in \mathtt{fst}(k`x)$.

PROPOSITION 4.4.2 *The* $\in_{\mathbf{co}}$ *of definition 4.4.1 is extensional.*

Proof: The proof is an inductive proof by cases. The situation we are contemplating is two distinct x_1 and x_2 which have the same $\in_{\mathbf{co}}$-members. Naturally we will be interested in the second components of $k`x_1$ and $k`x_2$. There are several cases to consider:

1. The second components are both $1_{\mathcal{S}}$, the unit of the semigroup. In this case the first components must be the same, and $x_1 = x_2$ follows.

2. One of the second components is $1_{\mathcal{S}}$ and the other isn't. Suppose $\mathtt{snd}(k`x_1) = 1_{\mathcal{S}}$ and $\mathtt{snd}(k`x_2)$ is $\mathtt{bw}$ or $\mathtt{cbw}$. (We cannot have two adjacent $\mathtt{c}$s since the words are reduced and $\mathtt{c}^2 = 1_{\mathcal{S}}$.) Then the collection of y such that $y \in_{\mathbf{co}} x_1$ is $\mathtt{fst}(k`x_1)$, which is a set in the sense of $\langle V, \in\rangle$. In contrast the collection of $y$ such that $y \in_{\mathbf{co}} x_2$ is either (i) the collection of $y$ such that $y \notin_{\mathbf{co}} k^{(-1)}\langle \mathtt{fst}(k`x_2), \mathtt{b(w)}\rangle$—which is to say, by an application of part 2 of definition 4.4.1—the same as the collection of y

such that $k^{(-1)}\langle \mathtt{fst}(k`x_2),\mathtt{w}\rangle \notin_{\mathbf{co}} y$, or (ii) the collection of $y$ such that $k^{(-1)}`\langle \mathtt{fst}(k`x_2), w\rangle \in_{\mathbf{co}} y$. The collection in case (i) cannot be a set because for any object a we can easily find proper-class-many unordered pairs which do not have a as a member. The collection in case (ii) cannot be a set because for any object a we can easily find proper-class-many unordered pairs which *do* have a as a member.

3. $\mathtt{snd}(k`x_1) = \mathtt{c}(\mathtt{w}_1)$ and $\mathtt{snd}(k`x_2) = \mathtt{c}(\mathtt{w}_2)$ where the first letter of both $\mathtt{w}_1$ and $\mathtt{w}_2$ is b. Assume $(\forall y)(y \in_{\mathbf{co}} x_1 \longleftrightarrow y \in_{\mathbf{co}} x_2)$ with a view to deducing $x_1 = x_2$. We can expand this in accordance with part 2 of definition 4.4.1 to get

$$(\forall y)(y \notin_{\mathbf{co}} k^{(-1)}\langle \mathtt{fst}(k`x_1),\mathtt{w}_1\rangle \longleftrightarrow y \notin_{\mathbf{co}} k^{(-1)}\langle \mathtt{fst}(k`x_2),\mathtt{w}_2\rangle)$$

which is to say

$$(\forall y)(y \in_{\mathbf{co}} k^{(-1)}\langle \mathtt{fst}(k`x_1),\mathtt{w}_1\rangle \longleftrightarrow y \in_{\mathbf{co}} k^{(-1)}\langle \mathtt{fst}(k`x_2),\mathtt{w}_2\rangle).$$

That is to say, if we have distinct x_1 and x_2 with the same members-in-the-sense-of-$\in_{\mathbf{co}}$, where the first letter of $\mathtt{snd}(k`x_1)$ is the same as the first letter of $\mathtt{snd}(k`x_2)$, namely c, then there are distinct y_1 and y_2 (namely $k^{(-1)}\langle \mathtt{fst}(k`x_1),\mathtt{w}_1\rangle$ and $k^{(-1)}\langle \mathtt{fst}(k`x_2),\mathtt{w}_2\rangle$) satisfying $(\forall z)(z \in_{\mathbf{co}} y_1 \longleftrightarrow z \in_{\mathbf{co}} y_2)$ and $\mathtt{snd}(k`y_1)$ and $\mathtt{snd}(k`y_2)$ are shorter than $\mathtt{snd}(k`x_1)$ and $\mathtt{snd}(k`x_2)$—indeed they are terminal segments.

This clearly reduces to the case where the two words in question are $\mathtt{w}_1$ and $\mathtt{w}_2$ both beginning with b, which we now treat.

4. The second components of $k`x_1$ and $k`x_2$ are words $\mathtt{b}(\mathtt{w}_1)$ and $\mathtt{b}(\mathtt{w}_2)$. We have $(\forall y)(y \in_{\mathbf{co}} x_1 \longleftrightarrow y \in_{\mathbf{co}} x_2)$. Expanding this by clause 2 in definition 4.4.1 we obtain

$$(\forall y)(k^{(-1)}`\langle \mathtt{fst}(k`x_1),\mathtt{b}(\mathtt{w}_1)\rangle \in_{\mathbf{co}} y \longleftrightarrow k^{(-1)}`\langle \mathtt{fst}(k`x_2),\mathtt{b}(\mathtt{w}_2)\rangle \in_{\mathbf{co}} y).$$

Now as long as

$$k^{(-1)}`\langle \mathtt{fst}(k`x_1),\mathtt{b}(\mathtt{w}_1)\rangle \neq k^{(-1)}`\langle \mathtt{fst}(k`x_2),\mathtt{b}(\mathtt{w}_2)\rangle$$

we can falsify this biconditional by taking y to be the singleton of one of these two. Singletons exist by low comprehension: the singleton of x (in the sense of $\in_{\mathbf{co}}$) is just $k^{(-1)}`\langle\{x\},0\rangle$.

$k`x_1$ has second component $\mathtt{cbw}_1$ and $k`x_2$ has second component $\mathtt{b}(\mathtt{w}_2)$. This is really the case where one word begins with a c and the other begins with a b but since the words are all reduced we also know that the letter following the c in the word beginning with a c must be a b. If we have

$$(\forall y)(y \in_{\mathbf{co}} x_1 \longleftrightarrow y \in_{\mathbf{co}} x_2)$$

this becomes

$$(\forall y)(y \notin_{\mathbf{co}} k^{(-1)}\langle \mathtt{fst}(k`x_1),\mathtt{b}(\mathtt{w}_1)\rangle \longleftrightarrow k^{(-1)}`\langle \mathtt{fst}(k`x_2),\mathtt{w}_2\rangle \in_{\mathbf{co}} y).$$

This is the same as

$(\forall y)(y \in_{\mathbf{co}} k^{(-1)}\langle \mathtt{fst}(k`x_1),\mathtt{b}(\mathtt{w}_1)\rangle \longleftrightarrow k^{(-1)}`\langle \mathtt{fst}(k`x_2),\mathtt{w}_2\rangle \notin_{\mathbf{co}} y)$.

Now $y \in_{\mathbf{co}} k^{(-1)}\langle \mathtt{fst}(k`x_1),\mathtt{b}(\mathtt{w}_1)\rangle$ iff (by clause 2 in definition 4.4.1)

$$k^{(-1)}\langle \mathtt{fst}(k`x_1),\mathtt{w}_1\rangle \in_{\mathbf{co}} y$$

and substituting this for '$y \in_{\mathbf{co}} k^{(-1)}\langle \mathtt{fst}(k`x_1),\mathtt{b}(\mathtt{w}_1)\rangle$' in

$$(\forall y)(y \in_{\mathbf{co}} k^{(-1)}\langle \mathtt{fst}(k`x_1),\mathtt{b}(\mathtt{w}_1)\rangle \longleftrightarrow k^{(-1)}`\langle \mathtt{fst}(k`x_2),\mathtt{w}_2\rangle \notin_{\mathbf{co}} y)$$

we obtain

$$(\forall y)(k^{(-1)}\langle \mathtt{fst}(k`x_1),\mathtt{w}_1\rangle \in_{\mathbf{co}} y \longleftrightarrow k^{(-1)}`\langle \mathtt{fst}(k`x_2),\mathtt{w}_2\rangle \notin_{\mathbf{co}} y).$$

Now anything like $(\forall y)(a \in y \longleftrightarrow b \notin y)$ must always be false because of the existence of the empty set. ■

The rest is easy:

PROPOSITION 4.4.3 *With* $\in_{\mathbf{co}}$ *as in definition 4.4.1,* $\langle V, \in_{\mathbf{co}}\rangle \models (\forall x)(-x$ *exists) and* $(\forall x)(B`x$ *exists).*

Proof: The complement of x will be $k^{(-1)}\langle \mathtt{fst}(k`x),\mathtt{csnd}(k`x)\rangle$, for, by clause 1 in definition 4.4.1, we have

$y \in_{\mathbf{co}} k^{(-1)}\langle \mathtt{fst}(k`x),\mathtt{csnd}(k`x)\rangle$ iff
$y \notin_{\mathbf{co}} k^{(-1)}\langle \mathtt{fst}(k`x), \mathtt{snd}(k`x)\rangle$ iff
$y \notin_{\mathbf{co}} x$.

$B`x$ will be $k^{(-1)}\langle \mathtt{fst}(k`x),\mathtt{bsnd}(k`x)\rangle$, for, by clause 2 in definition 4.4.1,

$y \in_{\mathbf{co}} k^{(-1)}\langle \mathtt{fst}(k`x),\mathtt{bsnd}(k`x)\rangle$ iff
$k^{(-1)}\langle \mathtt{fst}(k`x), \mathtt{snd}(k`x)\rangle \in_{\mathbf{co}} y$ iff
$x \in_{\mathbf{co}} y$.

■

4.4.2 *$\mathcal{P}$-extending models of Zermelo to models of NFO*

Before we contemplate the second construction (which gives us a model containing $x \cap y$ and $x \cup y$ for all x and y) we had better ask ourselves why we didn't get it last time. After all, these axioms hold in Oswald's model. The point is that in Oswald's model everything is finite or cofinite. Also if x and y are both finite or cofinite, then so are $x \cap y$ and $x \cup y$, and so $\cap$ and $\cup$ do not construct anything that isn't already there. Once we have $B`x$ this breaks down, and if we want $x \cup y$ and $x \cap y$ in general we have to construct them specially. This makes the second construction altogether more daunting.

THEOREM 4.4.4 *Every model of ZF + foundation has a $\mathcal{P}$-extension that is a model of NFO + low comprehension.*

Proof:

(We certainly need not restrict ourselves to models of ZF, for this is certainly true for well-founded models of Z and presumably weaker theories as well.)

The idea of the construction was originally that $k`x$ is to be a pair $\langle y, w\rangle$ where y is a set and w is a reduced word in some algebra with operations that correspond to the operations we want the universe of the new model to be closed under. Although this can be made to work, the approach it gives is very much less smooth than an approach that creates a gigantic free *NFO* model over a proper class of generators where there is one generator for each set of the old model. Unfortunately this second, smoother approach is not really a CO construction, and so strictly doesn't belong here as an illustration. The excuse for putting it in here is that the construction is, in spirit, very close to the CO constructions that precede it and will succeed it, and its presence here will help.

In the last case the only operations in which we were interested were unary (complementation and B) so the case had a spurious simplicity. Recall that the axioms of *NFO* are (apart from extensionality) ι (i.e. singleton), B, $\cup$, $\cap$, and complementation. If this were a CO construction *au pied de la lettre* we would get closure under ι free because of low comprehension, so we could explicitly forget about it. Here too we can forget about it, and will indeed do so—for the moment. Later we will see why this is all right after all. We should be able to make do with only one of $\cap$ and $\cup$, but conjunctive and disjunctive normal forms for boolean words are so useful that we will retain both.

Our language of terms has a constant term g_x for each old set x. The constants will eventually correspond to low sets of the new model. We also have function letters $\cup$, $\cap$, $-$, and B. We have to do a bit of work to find the correct notion of *reduced word* for this algebra. There is the irritating feature that we do not want to have both $a \cap b$ and $b \cap a$ but we can get round that by well-ordering the alphabet and extending the order lexicographically to the words. It will turn out that we will want to augment the language by adding Δ (symmetric difference).

Let us define a *restricted word* by recursion as follows: a restricted word is either a constant, or is $W \Delta g$ where W is a boolean combination of Bs of restricted words, and g is a constant. If g is missing we can speak of a *pure* restricted word.

Now we have to show that everything that we wish to construct can be denoted by a restricted word.

We can think of a word $\mathtt{w}$ as a boolean combination as a union of intersections, where the things being intersected are constants or Bs and complements of either. Consider an intersection like

$$a \cap b \cap c \cap d \ldots.$$

If even one of these is a constant (i.e. will correspond to a low set) then the whole intersection can be represented by just one (new) constant. Intersections of any number of complements of low sets can be represented as one complement of a low set. Thus the intersections are either constants, or intersections of Bs and $-B$s with the complement of a low set, which is to say, an intersection of Bs and $-B$s minus some low set.

Thus w can be rewritten in the form

$$[(w_1 \cap \overline{g_1}) \cup (w_2 \cap \overline{g_2}) \cup (w_3 \cap \overline{g_3}) \ldots] \cup g_{n+1}$$

where the w_i are intersections of values of B or complements of values of B, and the g_i are low sets. We will work through this in the case where $n = 3$, so that the reader can see how to do the general case. (This is probably more helpful than a rigorous proof of the general case would be!) For the moment we are interested only in the stuff inside the square brackets:

$$(w_1 \cap \overline{g_1}) \cup (w_2 \cap \overline{g_2}) \cup (w_3 \cap \overline{g_3}).$$

This expands to an intersection of $2^3 = 8$ subformulae as follows:

$(w_1 \cup w_2 \cup w_3) \cap$ seven other terms.

A typical example of these other terms is $w_1 \cup w_2 \cup \overline{g_3}$. It is typical in that it contains at least one entry of the kind $\overline{g_i}$. Now $w_1 \cup w_2 \cup \overline{g_3}$ is the same as

$$-(\overline{w_1} \cap \overline{w_2} \cap g_3).$$

This is the complement of a low set so we can think of this as $\overline{g_{\mathrm{novel}_1}}$ for some novel low set g_{novel_1}. This can be done to all the six remaining terms so the stuff-inside-the-square-brackets now looks like

$$(w_1 \cup w_2 \cup w_3) \cap \overline{g_{\mathrm{novel}_1}} \cap \overline{g_{\mathrm{novel}_2}} \cap \overline{g_{\mathrm{novel}_3}} \cap \ldots \cap \overline{g_{\mathrm{novel}_7}}.$$

Subtracting finitely many low sets is the same as subtracting one, so the stuff-inside-the-square-brackets is now

$$(w_1 \cup w_2 \cup w_3) \cap \overline{g_{\mathrm{novel}}}.$$

So, in the general case, we have reduced w to

$$(\bigcup_{i \leq n} w_n - g_{\mathrm{novel}}) \cup g_{n+1}.$$

This is actually

$$(\bigcup_{i \leq n} w_n) \Delta G$$

where G is

$$((\bigcup_{i\leq n} w_n) \cap g_{\text{novel}}) \cup (g_{n+1} \cap -(\bigcup_{i\leq n} w_n))$$

which is a low set.

Notice that we started with $\mathtt{w}$ as a boolean combination of Bs and constants and have ended up with something rather simpler: $\bigcup_{i\leq n} w_n$ is a boolean combination of Bs and G can be taken to be a constant.

This tidying-up process has not turned $\mathtt{w}$ into a restricted word, but if we invoke a notion of rank (where the rank of a word is simply the maximal depth of nesting of Bs in it) then we can see that none of the manipulations involved in the foregoing increases the rank of $\mathtt{w}$, so that if we perform these manipulations successively on words of increasing rank, every word will eventually be manipulated into a restricted word.

Every word is equivalent to a restricted word. Verifying extensionality in this model will depend on our being able to show that it is equivalent to a *unique* restricted word. (Note that it has not yet been made clear quite *how* it will depend on this uniqueness!) To do this it will be sufficient to show that if $\mathtt{w}_1$ and $\mathtt{w}_2$ are two pure restricted words then $\mathtt{w}_1 \,\Delta\, \mathtt{w}_2$ is not low. Suppose this were not so. Then there is some word W which evaluates to a low set. Without loss of generality we can think of W as a union of lots of terms each of which of course must themselves evaluate to something low and each such term is of the form $B`w_1 \cap B`w_2 \cap \ldots \cap B`w_n \cap -B`u_1 \cap -B`u_2 \cap \ldots \cap -B`u_k$ (where the $\vec{u}$ and the $\vec{w}$ are all distinct) which is supposed to be a low set. (Here $B`x$ means the set (in the sense of $\langle V, \in\rangle$) of those $y$ such that $x \in_{\mathbf{co}} y$.) Now $\{w_1, \ldots, w_n\} \cup \iota`x$ belongs to this collection as long as x is not a u or a w, and there is certainly going to be a proper class of such x so it is clear that this collection has to be a proper class (in the sense of $\langle V, \in\rangle$).

Now we have to define a new membership relation $\in_{\mathbf{co}}$. There is a bijection between the words of this new algebra and the sets of our old model. We may as well call this k as before. The elements of the model will be the words of the algebra. We will have to define $\in_{\mathbf{co}}$ by recursion on the structure of the terms.

If w is a molecular term (not a constant) then, in full generality, it is $w'\Delta g$ for some constant g and some boolean combination w' of $B`z_i$ for various i. We then say $x \in_{\mathbf{co}} w$ iff $(x \notin_{\mathbf{co}} g) \longleftrightarrow$(the obvious boolean combination of things like $z_i \in_{\mathbf{co}} x$). This is a recursive definition that appeals to a rank function on *NFO* terms, where rank is the depth of nesting of Bs. What are we to make of '$x \in_{\mathbf{co}} g_y$', where g_y is a constant? This is defined to be $k^{-1}`x \in y$. This ensures that every constant corresponds to a low set and vice versa. ■

One corollary of this is that any fragment of Z strong enough to execute this construction proves every $\Pi_1^{\mathcal{P}}$ theorem of *NFO*.

Before we proceed to other matters we should mention that the proof in Forster [1987b] that a theory there called *NF*$\forall$ had a recursive term model can probably be refined to show that every well-founded model of *ZF* has a $\mathcal{P}$-extension which is a model of *NF*$\forall$.

4.5 Church's model

Church's models are all rough CO constructions in which every hereditarily low set has an n-cardinal. What is an n-cardinal? The 1-cardinal of x is the cardinal of x in the usual sense. Actually it doesn't matter a great deal what happens for larger n, since whatever we decide to mean by it can be made to work. Sheridan is developing a construction that accommodates a version different from the one here (with the effect that the singleton function, considered as a set of Wiener–Kuratowski ordered pairs, is a union of finitely many of these cardinals). In the version used here, two sets have the same $(n+1)$-cardinal iff there is a bijection between them such that the elements paired by the bijection have the same n-cardinal. We start off with the case $n = 1$ for simplicity's sake.

As with the example of the last section, this is not a strict CO construction. We need a bijection (k, as ever) between V and a set of codes for objects. We will use the notation '$\overline{\overline{x}} = \overline{\overline{y}}$' to mean that x and y are the same size. We will see very soon how these fake terms ('$\overline{\overline{x}}$' etc.) can be treated as genuine denoting terms.

DEFINITION 4.5.1 *The things that are values of k are either*

1. *ordered pairs $\langle x, i\rangle$ where x is an arbitrary set and i is 0 or 1 (this will provide low sets and complements of low sets as usual) or*
2. *ordered pairs $\langle i, \kappa\rangle$ where κ is a cardinal (other than 0) and i is either I or II. I and II are two unspecified distinct objects that weren't in the original ground model. The idea is that these objects are to be cardinals (and complements of cardinals) in the new model.*

Now we say $y \in_{\mathbf{CO}} x$ iff

1. $\mathtt{snd}(k`x) = 0$ *and* $y \in \mathtt{fst}(k`x)$ *or*
2. $\mathtt{snd}(k`x) = 1$ *and* $y \notin \mathtt{fst}(k`x)$ *or*
3. $\mathtt{fst}(k`x) = I$ *and* $\mathtt{snd}(k`y) = 0$ *(so y is low) and* $\overline{\overline{\mathtt{fst}(k`y)}} = \mathtt{snd}(k`x)$ *or*
4. $\mathtt{fst}(k`x) = II$ *and* $(\mathtt{snd}(k`y) \neq 0$ *(so y is low) or* $\overline{\overline{\mathtt{fst}(k`y)}} \neq \mathtt{snd}(k`x))$.

Clauses 1 and 2 make sure that every low set has a complement. Notice that nothing has been said about what cardinal numbers *are*. Notice also that this does not matter! All we need is that there should be a definable class $\mathcal{C}$ and a definable relation belongs-to between sets and members of $\mathcal{C}$ satisfying

$$(\forall x)(\exists! y \in \mathcal{C})(x \ \mathtt{belongs\text{-}to} \ y)$$

$$(\forall x \forall y)(\overline{\overline{x}} = \overline{\overline{y}} \longleftrightarrow (\forall z \in \mathcal{C})(x \ \mathtt{belongs\text{-}to} \ z \longleftrightarrow y \ \mathtt{belongs\text{-}to} \ z)).$$

The term $\overline{\overline{x}}$ can then be taken to denote the appropriate member of $\mathcal{C}$. We do not need the axiom of choice to define cardinal numbers since as long as we have foundation (which we are assuming here) we can use Scott cardinals. The Scott cardinal of x is the set of all things the same size as x that are of minimal rank with this property.

Clause 3 will ensure that every low set has a cardinal in the new model (in the strong sense that for every low set x, the collection of all sets that have the same cardinal as x is a set of the new model). We stipulate that cardinals used do not include 0. We do this for two reasons: (i) to keep 0 free to signal low sets as usual, also (ii) because the extension of the cardinal number 0 (the set of all empty sets) is a set by low comprehension anyway, and we do not wish to make difficulties for ourselves with extensionality by manufacturing it twice. Clause 4 ensures that the complement of every such cardinal is a set.

The usual apparatus of low comprehension can now be taken for granted. It should by now be clear that this model is a model of complementation. It is the existence of cardinals that we had better spend a bit of time verifying.

PROPOSITION 4.5.2 *The clauses of definition 4.5.1 give a model in which every low set x has a cardinal:* $\{y : \overline{\overline{y}} = \overline{\overline{x}}\}$.

Proof: Notice that the cardinals that we have created by this means are demonstrably neither low nor are the complements of low sets, which makes life much easier. Let x be any low set. Consider the ordered pair $\langle \mathrm{I}, \overline{\overline{\mathtt{fst}(k\text{‘}x)}}\rangle$. We will check that $k^{(-1)}\text{‘}\langle \mathrm{I}, \overline{\overline{\mathtt{fst}(k\text{‘}x)}}\rangle$ is the cardinal of x (in the sense that it is the set of all things the same size as x) in the new model. By clause 3 we have $y \in_{\mathbf{co}} k^{(-1)}\text{‘}\langle \mathrm{I}, \overline{\overline{\mathtt{fst}(k\text{‘}x)}}\rangle$ iff y is low and $\overline{\overline{\mathtt{fst}(k\text{‘}y)}} = \overline{\overline{\mathtt{fst}(k\text{‘}x)}}$. Since x and y are both low this is the same as saying that the set (in the old sense) of things $\in_{\mathbf{co}} y$ is the same size (in the old sense) as the set of things $\in_{\mathbf{co}} x$, so there is a bijection between these two (old) sets. This bijection is an (old) set of (old) ordered pairs. By low comprehension (theorem 4.2.2) the corresponding (new) set of (new) ordered pairs is also a set, so x and y are of the same size in the new sense as well. The other direction is easy. Therefore $k^{(-1)}\text{‘}\langle \mathrm{I}, \overline{\overline{\mathtt{fst}(k\text{‘}x)}}\rangle$ is indeed the cardinal of x in the new model. Correspondingly $k^{(-1)}\text{‘}\langle \mathrm{II}, \overline{\overline{\mathtt{fst}(k\text{‘}x)}}\rangle$ is the complement of that cardinal, which we have to have if complementation is to be true. ■

We have to do a little bit of work to see how to generalize this correctly to the case $n = 2$, the model where every low set of low sets has a 2-cardinal. What Church actually claims is that for each n his construction gives us a model where every well-founded set has an n-cardinal. I prefer the statement in terms of low sets, low sets of $\ldots^n$ low sets.

For the case $n = 2$ we have to add two more clauses 5 and 6 to definition 4.5.1 in the same style. We will need two more novel constants in the style of I and II, which we may as well write 'III' and 'IV'. Objects x s.t. $\mathtt{fst}(k`x) =$ III will be 2-cardinals and objects $x$ s.t. $\mathtt{fst}(k`x) =$ IV will be complements of 2-cardinals. We will need the notation '$\mathtt{2\text{-}card}`x$' for the 2-cardinal of x, and we will use lower case Greek letters to range over 2-cardinals as over cardinals. Then there are to be two further kinds of ordered pairs in the range of k: pairs whose first components are III and pairs whose first components are IV. In both cases the second components are 2-cardinals. We will need the two following new clauses in the definition of $y \in_{\mathbf{co}} x$.

DEFINITION 4.5.3
5: $\mathtt{fst}(k`x) =$ III *and* $y$ *is a low set of low sets and* $\mathtt{snd}(k`x) = \mathtt{2\text{-}card}`\{\mathtt{fst}(k`z) : z \in \mathtt{fst}(k`y)\}$
6: $\mathtt{fst}(k`x) =$ IV *and* ($y$ *is not a low set of low sets or* $\mathtt{snd}(k`x) \neq 2 - \mathtt{card}`\{\mathtt{fst}(k`z) \colon z \in \mathtt{fst}(k`y)\}$).

Clause 6 of course ensures that 2-cardinals, too, have complements. The details will be omitted.

PROPOSITION 4.5.4 *The membership relation of definition 4.5.3 gives a model in which each low set of low sets has a 2-cardinal.*

Proof: Let x be a low set of low sets. Then the 2-cardinal (in the sense of $\in_{\mathbf{co}}$) of x will be $k^{(-1)}`\langle \text{III}, \mathtt{2\text{-}card}`\{\mathtt{fst}(k`z) : z \in \mathtt{fst}(k`x)\}\rangle$. We had better check this. Suppose y is a low set of low sets. Then

$$y \in_{\mathbf{co}} k^{(-1)}`\langle \text{III}, \mathtt{2\text{-}card}`\{\mathtt{fst}(k`z) : z \in \mathtt{fst}(k`x)\}\rangle$$

iff

$$\mathtt{2\text{-}card}`\{\mathtt{fst}(k`z) : z \in \mathtt{fst}(k`x)\} = \mathtt{2\text{-}card}`\{\mathtt{fst}(k`z) : z \in \mathtt{fst}(k`y)\}.$$

What we actually want is for x and y to have the same 2-cardinal in the new sense. As before, if there is an (old) bijection between $\mathtt{fst}(k`x)$ and $\mathtt{fst}(k`y)$ there will be a new bijection between x and y by low comprehension. And the same goes, not only for x and y, but for each $x' \in_{\mathbf{co}} x$ and $y' \in_{\mathbf{co}} y$ that are paired by the bijection: if there is an (old) bijection between $\mathtt{fst}(k`x')$ and $\mathtt{fst}(k`y')$ there will be a new bijection between x' and y' by low comprehension as desired. ■

It should now be clear how to tinker with this construction to add n-cardinals simultaneously for all $n \in \mathbb{N}$.

4.6 Mitchell's set theory

Sadly, Mitchell [1976] has never been published. Mitchell set theory is extensionality, complementation, power set, replacement for well-founded sets, and the assertion that $WF \models ZF$.

Mitchell defines a class A, with a binary relation E over it (I shall stick to his notation), so that $\langle A, E\rangle$ is a model of his theory, and his presentation is in the style (ii) of the list at the beginning of this chapter. A will be the set of quadruples of a particular kind. To be precise, he defines a predicate $A_{i,j,\alpha}(x)$ by the following recursion:

False unless i and j are integers with $i = 0, 1, 2$, or 3, with α an ordinal.
False if $j = 0$ and $i \neq 0$ or 1.
False if $j > 0$ and $i = 0$ or 1.
$A_{0,0,0}(x)$ iff x is a quadruple whose last three components are all 0.
$A_{1,0,0}(x)$ iff x is a quadruple whose last three components are $1, 0, 0$.
$A_{2,j+1,\alpha}(x)$ iff $(\exists y)(\exists i \in \{1,2,3\})A_{i,j,\alpha}(y) \wedge x = \langle y, 2, j+1, \alpha\rangle \wedge y \neq \langle 0,1,0,0\rangle$.
$A_{3,j+1,\alpha}(x)$ iff $(\exists y)(\exists i \in \{1,2,3\})A_{i,j,\alpha}(y) \wedge x = \langle y, 3, j+1, \alpha\rangle \wedge y \neq \langle 0,1,0,0\rangle$.
For $\alpha > 0$, $A_{0,0,\alpha}$ iff $(\exists y)(x = \langle y, 0, 0, \alpha\rangle \wedge (\forall z \in y)\exists i \exists j(\exists \beta < \alpha)A_{i,j,\beta}(z)$ and the l.u.b. of $\{\beta+1 : \exists i \exists j(\exists z \in y)(A_{i,j,z}(z) \wedge \neg A_{0,0,0}(z))\} = \alpha)$.
For $\alpha > 0$, $A_{1,0,\alpha}$ iff $\exists y[x = \langle y, 1, 0, \alpha\rangle \wedge A_{0,0,\alpha}(\langle y, 0, 0, \alpha\rangle)]$.

Things x such that $(\exists ij\alpha)A_{i,j,\alpha}(x)$ are *A-objects.* The complement $\overline{x}$ of an A-object x is the result of swapping 0 for 1 (and vice versa) and 3 for 2 (and vice versa) in the second coordinate of x. If the second coordinate of x is 1 or 3 then x is a *complement* object; if it is 0 x is a *replacement object* and, if it is 2, x is a *power object.* $\overline{x}$ will be the complement of x in the sense of the new model. If $x = \langle x', 0, j_x, \alpha_x\rangle$ is a replacement object, let m_x be x' if $\alpha_x > 0$, and $\{y : A_{0,0,0}(y) \wedge y' \in x'\}$ otherwise (y' is the first coordinate of y). m_x is the set of A-objects which will be members of x in the new sense. The last auxiliary function we will need is P which will give us *power* objects. Let x be $\langle x', i_x, j_x, \alpha_x\rangle$; then

$P`x = x$ if $x = \langle 0,1,0,0\rangle$
$P`x = \langle x, 2, j_x + 1, \alpha_x\rangle$ if $i_x = 1, 2$, or 3
$P`x = \langle \{z : z \subseteq x'\}, 0, 0, 0\rangle$ if $i_x = 0$ and $\alpha_x = 0$
$P`x = \langle \{z : i_z = 0 \wedge \alpha_z \subseteq \alpha_x \wedge m_z \subseteq m_x\}, 0, 0, \alpha_x\rangle$ if $i_x = 0$ and $\alpha_x > 0$.

P is intended to be the power set operation of the new model, and we have to define E recursively in such a way that this is the case. The lack of the axiom of foundation is a serious problem here, for it can happen that an object can have a proper class of members, and therefore we keep having to add new members. The particular problem here arises because of the power set operation, and can be circumvented by a simultaneous recursion which defines E and a "subset" relation S at the same time.

Only A-objects can be in the domain of S and E, and the other clauses in the recursion are as follows:

If $A_{0,0,0}(y)$ then $(\forall x)(x\ E\ y \longleftrightarrow A_{0,0,0}(x) \wedge x' \in y')$
If $A_{0,0,0}(y)$ then $(\forall x)(x\ S\ y \longleftrightarrow A_{0,0,0}(x) \wedge x' \subseteq y')$
$\alpha > 0 \rightarrow A_{0,0,\alpha}(y) \rightarrow (x\ E\ y \longleftrightarrow A(x) \wedge x \in y')$
$\alpha > 0 \rightarrow A_{0,0,\alpha}(y) \rightarrow (x\ S\ y \longleftrightarrow i_x = 0 \wedge m_x \subseteq m_y)$
$A_{1,0,\alpha}(y) \rightarrow (x\ E\ y \longleftrightarrow \neg(x\ E\ \overline{y}))$
$A_{2,j,\alpha}(y) \rightarrow (x\ E\ y \longleftrightarrow x\ S\ y')$
$A_{3,j,\alpha}(y) \rightarrow (x\ E\ y \longleftrightarrow \neg(x\ E\ \overline{y}))$
$A_{1,0,\alpha}(y) \rightarrow A_{0,0,\beta}(x) \rightarrow (x\ S\ y \longleftrightarrow m_x \cap m_{\overline{y}} = \Lambda)$
$A_{1,0,\alpha}(y) \rightarrow A_{1,0,\beta}(x) \rightarrow (x\ S\ y \longleftrightarrow m_{\overline{y}} \subseteq m_{\overline{x}})$
$A_{1,0,\alpha}(y) \rightarrow A_{2,j,\beta}(x) \rightarrow (x\ S\ y \longleftrightarrow \forall w \in m_{\overline{y}}\ \neg w\ E\ x)$
$A_{1,0,\alpha}(y) \rightarrow A_{3,j,\beta}(x) \rightarrow (x\ S\ y \longleftrightarrow \overline{y}\ S\ \overline{x})$
$A_{2,j,\alpha}(y) \rightarrow (A_{1,0,\beta}(x) \vee A_{3,k,\delta}(x) \rightarrow (x\ S\ y \longleftrightarrow x \neq x))$
$A_{2,j,\alpha}(y) \rightarrow (A_{0,0,\beta}(x) \rightarrow (x\ S\ y \longleftrightarrow (\forall w \in m_{\overline{x}})\ w\ S\ y'))$
$A_{2,j,\alpha}(y) \rightarrow (A_{2,k,\beta}(x) \rightarrow (x\ S\ y \longleftrightarrow x'\ S\ y'))$
$A_{3,j,\alpha}(y) \rightarrow (A_{0,0,\beta}(x) \rightarrow (x\ S\ y \longleftrightarrow (\forall w \in m_{\overline{x}})\ \neg w E \overline{y}))$
$A_{3,j,\alpha}(y) \rightarrow (A_{1,0,\beta}(x) \vee A_{2,k,\delta}(x) \rightarrow (x\ S\ y \longleftrightarrow x \neq x))$
$A_{3,j,\alpha}(y) \rightarrow (A_{3,k,\beta}(x) \rightarrow (xSy \longleftrightarrow \overline{y}\ S\ \overline{x}))$.

The details of the proof that this is indeed a model of Mitchell set theory can be found in Mitchell [1976]. It seems to be an open question whether models of Mitchell set theory obtained in this way are automatically models of pseudofoundation. Mitchell has written[51] that he thinks that a counterexample could be found among the complements of power sets of complements of replacement objects.

4.7 Conclusions

Church–Oswald constructions can become very complicated and can be presented in various different ways. At least two of the constructions here are not strict CO constructions in the sense of the characterization in section 4.1. Strictly the use of terms 'I', 'II', 'III', and 'IV' is just a little bit naughty, since they do not denote anything in the model we start with. There are ways round this, but they involve the use of coding functions that make things even more messy. Sheridan starts with a model with a proper class of *urelemente* in the belief that this makes certain aspects of the exposition clearer. Swings and roundabouts. One thing that no attention to presentation can conceal is the essential triviality of the axiomatic systems whose consistency is being proved. The machinery just is very weak: none of the coding machinery seen so far will give us a model of the axiom of

[51] Personal communication.

infinity unless we started with one. This is because unless we can show that $\in_{\mathbf{CO}}$ is extensional we do not have a model of a set theory at all, and proving that $\in_{\mathbf{CO}}$ is extensional depends on the word problem of the theory (whose consistency we are attempting to prove) being easily solvable. Such theories are going to be very weak. To labour the point one can remark that the techniques developed here give no clue how to answer (for example) the question "Must every well-founded model of Zermelo have a $\mathcal{P}$-extension that is a model of *NF*?" *even on the assumption that NF is consistent.* None of the CO theories have the great and distinctive advantage that one might expect theories with a universal set to have, namely the natural—one might almost say naïve—treatment of inductive definitions. One cannot say that a desired datatype is the intersection of the set Y of all x containing this and closed under that, because the collection Y is never a set.

The moral is that what we get from this method of constructing models is a way of proving theorems of the following kind.

> *If T is a weak theory for which the word problem is tractable, then any model $\mathcal{M}$ of a weak well-founded set theory has a $\mathcal{P}$-extension that is a model of $T+$ low comprehension.*

In his seminal paper on *NFU*, Jensen shows how to make a model of *NFU* whose well-founded part is as big as desired. Interested readers should consult Jensen [1969]. Boffa, by using the methods of Grishin [1969], has shown that

REMARK 4.7.1 *Let T be a set theory. Expand NF_3 with a new constant A, and add axioms to say that $A \subseteq \mathcal{P}`A$ and that $A \models T$ by means of new variables restricted to A. Thus T is naturally interpreted in the new theory. Extend the comprehension scheme of NF_3 to the new language, with the liberalization that only the old variables need stratification. Then the new theory is a conservative extension of T.*

Boffa presented this at the Oberwolfach meeting of 1987 but, as far as I know, has not published a proof. The result in Hinnion [1990] has the same flavour.

Initially at least one should avoid overstating the case for the Church–Oswald technique and simply say that in skilled hands it can turn a model of *ZF* into a model of a set theory with $V \in V$ plus axioms saying that V is closed under whichever operations the hands' owner has decided to code. However, for a lot of the operations f for which it is easy to arrange for the universe to be closed under f it is also easy to persuade oneself that there is some philosophical justification for the assertion that the universe is closed under f. For example, Sheridan [199?] on the axiom of separation restricted to well-founded sets "The axiom of separation can be justified only for well-founded sets. One justifies separation by pretending that sets are located in time, as in type theory. *After* one has constructed some sets, one can construct a new set which contains those of the sets satisfying some open sentence .... If all sets arise in this way, then if one has a set, one could have constructed the set leaving out those not satisfying some open sentence. Hence the axiom of separation." This might be mere coincidence, and sceptical souls may well feel that they are merely hearing the grating sound of a virtue being made of necessity, but it just might be generally true that those axioms that can be modelled by this technique are those for which there is some philosophical justification. The question of the relationship between philosophical justification of axioms for set theory with a universal set and the availability of consistency proofs by this method is a good one, and it certainly deserves more attention from philosophers of mathematics than it has had so far.

We conclude with some problems. The first is to axiomatize the theories that CO constructions give us: low comprehension is very important but not obviously axiomatizable. Second, as Richard Kaye has pointed out to me, it would be interesting to know what constraints have to be put on the coding function k of the CO construction to ensure that pseudofoundation holds in the result. The last problem is the very classical nature of CO constructions: $\mathtt{snd}(k`x)$ is always equal to 0, or to something else. This reliance on the *tertium non datur* leaves it very unclear what an intuitionistic version would look like. There is a double-negation interpretation of classical *ZF* into intuitionistic *ZF* (due to Powell [1975]) which relies on constructing by $\in$-recursion an inner model of hereditarily well-behaved sets. Of course this cannot be done in a set theory with a universal set and it is an open problem whether or not there is a double-negation interpretation for such theories at all. (Indeed, as Dzierzgowski has shown, it is extraordinarily difficult to get non-trivial models for intuitionistic versions even of systems as straightforward as NF_2.) It would be quite illuminating to develop CO constructions inside intuitionistic *ZF* and see what happens to the Powell double-negation interpretation. Another question that arises naturally in this context is the following: can we show that no (internal) recursive construction like this over a model of *ZF* will produce a model of *NF*? Kaye has conjectured that there is no $\Delta_1^{\mathcal{P}}$ interpretation of *NF* over the well-founded sets of a model of *NF*.

5

OPEN PROBLEMS

5.1 Permutation models and quantifier hierarchies

CONJECTURE 5.1.1 *Every $\forall^1\exists^*$ sentence refutable in NF is refutable already in NF_2.*

CONJECTURE 5.1.2 *Every $\forall^*\exists^*$ sentence refutable in NF is refutable already in NFO.*

CONJECTURE 5.1.3 *NFO decides all stratified $\forall^*\exists^*$ sentences.*

CONJECTURE 5.1.4 *Any term model for NF and any model for NF in which all sets are symmetric satisfies every $\forall^*\exists^*$ sentence consistent with NFO.*

CONJECTURE 5.1.5 *All unstratified $\forall^*\exists^*$ sentences are either decided by NF or can be proved consistent by permutations.*

In particular, find permutation models of
Every self-membered set is infinite.

Friederike Körner [1994] has partially solved this with theorem 3.1.37. Every model of the sort provided by this theorem has a permutation model in which every self-membered set has a countable partition.

CONJECTURE 5.1.6 (Boffa). *Every self-membered set is the same size as V.*

Boffa conjectures that this is consistent with *NF*. The dual of this is $(\forall x)(x \notin x \to \overline{\overline{-x}}=\overline{\overline{V}})$. Now if these two hold simultaneously we infer $(\forall x)(\overline{\overline{x}} = \overline{\overline{V}} \vee \overline{\overline{-x}} = \overline{\overline{V}})$. This is certainly not going to be provably consistent by means of permutations. All that Bernstein's lemma tells us is that $(\forall x)(\overline{\overline{x}} = \overline{\overline{V}} \vee \overline{\overline{-x}} \geq_* \overline{\overline{V}})$. For this reason I find a more plausible conjecture to be that $NF \vdash \Diamond(\forall x)(x \in x \to \overline{\overline{x}} \geq^* \overline{\overline{V}})$.
Every self-membered set has a member that is not self-membered.
No self-membered set belongs to all its members.
There are no non-trivial $\in$-automorphisms of the universe. Since there is an automorphism in V^σ iff $(\exists\pi)(\sigma\pi\sigma^{-1} = j`\pi)$ it will be sufficient to find $\sigma$ such that $(\forall\pi)(\sigma\pi\sigma^{-1} \neq j`\pi)$.

Finally, find a sensible topology for the family of permutation models. By enabling us to use topological notions of "large" and "small" it might enable us to resolve questions like: "Which formulae of the form $\Diamond\phi$ are theorems of *NF*?"

5.2 Cardinals and ordinals in *NF*

What is the cofinality of the ordinals? Remark 2.2.12 tells us that the order type of $\langle NO, \leq \rangle$ is a successor initial ordinal (assuming AC_{wo} well-ordered choice) and so must be regular, but nothing is known without this assumption.

Does NF prove that there are infinitely many infinite cardinals? This is an old question of Specker's, and has resisted protracted assaults by Forster and Pétry. At present it still seems to be consistent with *NF* that NCI, the set of infinite cardinals, should be finite. The strongest known consequences of the assertion that NCI is finite is that NCI would be a complete distributive lattice, and $(\forall\alpha \in NCI)(\alpha = 2 \cdot \alpha)$.

Is the existence of a cardinal of infinite rank consistent with ZF? It is certainly not consistent with ZFC (proposition 2.2.13). Since we do not know how to prove NCI infinite, *a fortiori* we do not know how to prove the existence of cardinals of infinite rank in *NF* either. In NFC it is easy—$\overline{\overline{V}}$ is of non-cantorian (and therefore infinite) rank. If the consistency of cardinals of infinite rank should turn out to involve large cardinals (as some suspect) this would be a hint that NFC is *much* stronger than *NF*.

Does $AxCount_{\leq}$ *imply that for all countable ordinals* α, $\alpha \leq T\alpha$?

5.3 KF

Recall that $\exists NO$ is the assertion that there is a set X of well-orderings with the property that every well-ordering is the same length as some member of X.

Is KF + $\exists NO$ *consistent?*

Does KF + $\exists NO$ *refute the axiom of foundation?*

Are $KFI + \exists NO$ *and Z equiconsistent?*

Are $KFI + \exists NO$ *and NF equiconsistent?*

No consistency result is known for any set theory proving $\exists NO$. There is evidently a real consistency cost to having "universal" objects of this kind. *NF* takes the "same" approach (V and NO are both sets) to the two paradoxes of Cantor and Burali-Forti. *ZF* too provides the same answer to these paradoxes—though of course it is a different answer from that given by *NF*. (We do not seem to have much choice about our approach to Mirimanoff's and Russell's paradoxes—it is very difficult to arrange for either the Russell class or the collection of well-founded sets to be sets!) Do the answers to these two questions have to be linked? KF is a natural context in which to ask if these two decisions of *NF* have the same consistency strength. If so, then $KFI + \exists NO$ should be equiconsistent with *NF*.

The constructible model of KF As we have seen (theorem 2.3.26) there is a finite collection of stratified (but not homogeneous) Δ_0^{Levy} functions with the property that any model of power set and sumset that is closed under all of them is also a model of stratified Δ_0^{Levy} separation. This means that we can use them to perform (in, say, Kripke–Platek set theory) a construction analogous to the construction of L. This structure will presumably be a model for stratified fragments of *ZF* such as KF. It has a canonical enumeration (just as L does) but since this enumeration is highly unstratified there is no reason to suppose it can be coded inside our new structure, so there is no reason to suppose we have built a model of *AC*! The study of this structure has not been seriously begun. It certainly looks like a potentially very fruitful approach to KF.

One way of raising this question is a way that makes sense to *ZF*istes is to ask: does the smallest transitive set containing V_ω and closed under the stratified rudimentary functions contain a well-ordering of V_ω?

5.4 Other subsystems

Does NF∀ have a recursive term model?* It can be shown that *NF*∀ has a recursive term model (Forster [1987b]) but this is hard work and depends heavily on the finite axiomatizability of *NF*∀. On the other hand the things that would prevent this (definability of the naturals and so on) seem to need stronger comprehension axioms—specifically the existence of $\{x : \Phi\}$ where Φ has alternation of quantifiers. Since we cannot expect clean results once the integers become definable, this would be the best possible.

5.5 Well-founded extensional relations

In *NF* can there be a well-founded extensional relation on the universe that is a *set*? This is an old question of Hinnion's. The existence of a definable such relation would enable us to show (by induction on it) that any automorphism of $\langle V, \in\rangle$ must be the identity. We also know (proposition 3.1.10) that if $NF + AC_2$ is consistent, so is $NF + AC_2+$ there exists a non-trivial automorphism of $\langle V, \in\rangle$. Since the assertion that there is a well-founded extensional relation on V is stratified, and false in at least one permutation model of any model of $NF + AC_2$, it must be false in every model of $NF + \mathrm{AC}_2$, which is to say, $NF + AC_2 \vdash$ there is no well-founded extensional relation on V. Since AC_2 is false in any term model for *NF* there seems to be a possibility that there might be a formula ϕ such that in some models $\{\langle x, y\rangle : \phi\}$ is a well-founded extensional relation on V. Kaye has recently remarked that a model of KF containing a *transitive* x the same size as $\mathcal{P}`x$ gives rise to a model of *NF*+ there is a well-founded extensional relation on V. There is an old result of Rubin's that (assuming foundation) if the power set of a well-ordered set can always be well-ordered, the AC follows. In the NF context this becomes: if there is a well-founded extensional relation on the universe then the power set of a well-ordered set cannot always be well-ordered. Holmes has asked if it is consistent with NF that the domain of a well-founded extensional relation can always be well-ordered.

5.6 Term models

It is not known if NF can have a term model, nor even if it can have a model in which every set is symmetric. It may be that the existence of a term model for NF is stronger than the mere consistency of NF, and that if NF has a term model so does NFC (though beware: a term model for NF is not the same as an ω-model, for there might be closed terms potentially denoting non-standard integers: one possible candidate is "the cardinal number of the set of alephs"). Any term model is rigid, and consists of highly symmetric objects, so AC fails badly (AC_2 fails, for example: see section 2.1.3). Can DC be true in a term model for NF? Is it consistent with NF at all?

AxCount$_\leq$ is independent of NF if NF is consistent (theorem 3.1.34) and is equivalent to $\Diamond\exists\{V_n : n \in \mathbb{N}\}$ (theorem 3.1.28), so it implies that $\Diamond(V_\omega$ exists). It is easy to show that V_ω is not the extension of any closed stratified term. Can we get rid of V_ω by permutations? If not, consider a term model of $NF+$ AxCount$_\leq$. It has a permutation model¡ containing V_ω and must therefore itself contain V_ω, and therefore is not a term model. So $NF+$AxCount$_\leq$ would have no term model even if NF does. This does not seem very likely. On the other hand there is no obvious uniform way of producing a permutation to satisfy $\neg\exists V_\omega$. So far we have never proved $\Diamond\phi$ without being able to prove ϕ^τ for some *definable* τ. Of course it might turn out that $\Diamond\neg\exists V_\omega$ is consistent and independent, but so far $\Diamond\phi$ has always turned out to be provable or refutable or equivalent to ϕ (except where some artifice such as a Rosser sentence is involved). There remains the rather extreme formula AxCount$_\leq \longleftrightarrow \Box(WF \prec_{str(\Sigma_1^{\mathcal{P}})} V)$ on page 114 which has not yet been refuted.

5.7 Miscellaneous

Is it consistent with NF that every set should have a transitive closure? (Every set has a transitive *superset* because of V.) In any term model every set has a transitive closure because of proposition 2.1.8.
Does $Con(NF)$ follow from the existence of a model of NFU with an automorphism group acting transitively on distinct tuples of atoms?

Is there a double-negation interpretation of classical NF into intuitionistic NF?

$TNT \vdash Amb(\Sigma_1^{Levy})$?

Is Amb^n as strong as Amb?
$TNT \vdash Amb(\forall_2)$?
Is there a cut-free proof of the axiom of infinity?
Is Boffa's **W** (page 56) equal to V?

BIBLIOGRAPHY

This bibliography has been built on the basis of "if in doubt put it in" and it properly extends all previous bibliographies on this subject. With a reading list this long it is perverse not to give the reader some clues as to what repays study and what does not. There are many ways in which an item can come to be in this list. Several (and these are listed separately) are mentioned not because they are "about" set theory with a universal set, but merely because they happened to be mentioned in the body of the text; a large number (including some I have not read myself) are included simply in order that I shall not be accused of leaving anything out; finally there are the items that should be part of the further reading of anyone who wishes to pursue this subject seriously. Usually it will be clear from the context whether or not a citation of an article is an invitation to read it, but I have taken the precaution of sprinkling asterisks around to ensure that any articles I consider worth reading that may not already have such an obvious recommendation nevertheless leap to the eye. It should be pointed out here, for only a close reading of the bibliography would reveal it, that a large concentration of good articles on *NF* is to be found in *Cahiers du Centre de Logique* volume 4, published by CABAY, an arm of the Université Catholique de Louvain.

Concerning set theory with a universal set

Aczel, P. [1988] Non-well-founded sets. CSLI lecture notes, Stanford University (distributed by Chicago University Press).

Arruda, A. [1970a] Sur les systèmes NF_i de Da Costa. *Comptes Rendus hebdomadaires des séances de l'Académie des Sciences de Paris série A* **270** pp. 1081–4.

Arruda, A. [1970b] Sur les systèmes NF_ω. *Comptes Rendus hebdomadaires des séances de l'Académie des Sciences de Paris série A* **270** pp. 1137–9.

Arruda, A. [1971] La mathématique classique dans NF_ω. *Comptes Rendus hebdomadaires des séances de l'Académie des Sciences de Paris série A* **272** p. 1152.

Arruda, A. and Da Costa, N.C.A. [1964] Sur une hiérarchie de systèmes formels. *Comptes Rendus hebdomadaires des séances de l'Académie des Sciences de Paris série A* **259** pp. 2943–5.

Barwise, J. [1984] Situations, sets and the axiom of foundation. In *Logic Colloquium '84*. Ed. J. Paris, A. Wilkie, and G. Wilmers, North-Holland, Amsterdam, 1986 pp. 21-36.

Beneš, V.E. [1954] A partial model for NF. *Journal of Symbolic Logic* **19** pp. 197–200.

* Boffa, M. [1971] Stratified formulas in Zermelo–Fränkel set theory. *Bulletin de l'Académie Polonaise des Sciences série Math.* **19** pp. 275–80.

* Boffa, M. [1973] Entre NF et *NFU*. *Comptes Rendus hebdomadaires des séances de l'Académie des Sciences de Paris série A* **277** pp. 821–2.

* Boffa, M. [1975a] Sets equipollent to their power sets in NF. *Journal of Symbolic Logic* **40** pp. 149–50.

* Boffa, M. [1975b] On the axiomatization of NF. *Colloque international de Logique*, Clermont-Ferrand 1975. pp. 157–9.

* Boffa, M. [1977a] A reduction of the theory of types. *Set theory and hierarchy theory*, Springer Lecture Notes in Mathematics **619** pp. 95–100.

* Boffa, M. [1977b] The consistency problem for NF. *Journal of Symbolic Logic* **42** pp. 215–20.

* Boffa, M. [1977c] Modèles cumulatifs de la théorie des types. *Publications du Département de Mathématiques de l'Université de Lyon* **14** fasc 2. pp. 9–12.

* Boffa, M. [1981] La théorie des types et NF. *Bulletin de la Société Mathématique de Belgique série A* **33** pp. 21–31.

* Boffa, M. [1982] Algèbres de Boole atomiques et modelès de la théorie des types. *Cahiers du Centre de Logique* (Louvain-la-neuve) **4** pp. 1–5.

* Boffa, M. [1984a] Arithmetic and the theory of types. *Journal of Symbolic Logic* **49** pp. 621–4.

* Boffa, M. [1984b] The point on (*sic*) Quine's NF (with a bibliography). *TEORIA* **4** (fasc. 2) pp. 3–13.

* Boffa, M. [1988] ZFJ and the consistency problem for NF. *Jahrbuch der Kurt Gödel Gesellschaft* (Wien) pp. 102–6.

* Boffa, M. and Casalegno, P. [1985] The consistency of some 4-stratified subsystem of *NF* including NF_3. *Journal of Symbolic Logic* **50** pp. 407–11.

* Boffa, M. and Crabbé, M. [1975] Les théorèmes 3-stratifiés de NF_3. *Comptes Rendus hebdomadaires des séances de l'Académie des Sciences de Paris série A* **280** pp. 1657–8.

* Boffa, M. and Pétry, A. On self-membered sets in Quine's set theory *NF*. *Travaux de Mathématiques,* Université Libre de Bruxelles, Fascicule I pp. 25–6.

Church, A. [1974] Set theory with a universal set. *Proceedings of the Tarski Symposium.* Proceedings of Symposia in Pure Mathematics XXV, ed. L. Henkin, Providence, RI, pp. 297–308. Also in *International Logic Review* **15** pp. 11–23.

Cocchiarella, N.B. [1976] A note on the definition of identity in Quine's New Foundations. *Zeitschrift für mathematische Logik und Grundlagen der Mathematik* **22** pp. 195–7.

* Coret, J. [1964] Formules stratifiées et axiome de fondation. *Comptes Rendus hebdomadaires des séances de l'Académie des Sciences de Paris série A* **264** pp. 809–12 and 837–9.

* Coret, J. [1970] Sur les cas stratifiés du schema de remplacement. *Comptes Rendus hebdomadaires des séances de l'Académie des Sciences de Paris série A* **271** pp. 57–60.

* Crabbé, M. [1975] Types ambigus. *Comptes Rendus hebdomadaires des séances de l'Académie des Sciences de Paris série A* **280** pp. 1–2.

* Crabbé, M. [1976] La prédicativité dans les théories élémentaires. *Logique et Analyse* **74-75-76** pp. 255–66.

* Crabbé, M. [1978a] Ramification et prédicativité. *Logique et Analyse* **84** pp. 399–419.

* Crabbé, M. [1978b] Ambiguity and stratification. *Fundamenta Mathematicae* **CI** pp. 11–17.

* Crabbé, M. [1982a] On the consistency of an impredicative subsystem of Quine's NF. *Journal of Symbolic Logic* **47** pp. 131–6.

* Crabbé, M. [1982b] À propos de 2^{α}. *Cahiers du Centre de Logique* (Louvain-la-neuve) **4** pp. 17–22.

* Crabbé, M. [1983] On the reduction of type theory. *Zeitschrift für mathematische Logik und Grundlagen der Mathematik* **29** pp. 235–7.

* Crabbé, M. [1984] Typical ambiguity and the axiom of choice. *Journal of Symbolic Logic* **49** pp. 1074–8.

* Crabbé, M. [1986] Le schéma d'ambiguïté en théorie des types. *Bulletin de la Société Mathématique de Belgique série B* **38** pp. 46–57.

* Crabbé, M. [1991] Stratification and cut-elimination. *Journal of Symbolic Logic* **56** pp. 213–26.

* Crabbé, M. [1992] On *NFU*. *Notre Dame Journal of Formal Logic* **33** pp. 112–19.

* Crabbé, M. [1994] The Hauptsatz for stratified comprehension: a semantic proof. *Mathematical Logic Quarterly* **40** pp 481–9.

Curry, H.B. [1954] Review of Rosser [1953a]. *Bulletin of the American Mathematical Society* **60** pp. 266–272.

Da Costa, N.C.A. [1964] Sur une système inconsistent de théorie des ensembles. *Comptes Rendus hebdomadaires des séances de l'Académie des Sciences de Paris série A* **258** pp. 3144–7.

Da Costa, N.C.A. [1965a] Sur les systèmes formels C_i,C_i^*,$C_i^=$,D_i et *NF*. *Comptes Rendus hebdomadaires des séances de l'Académie des Sciences de Paris série A* **260** pp. 5427–30.

Da Costa, N.C.A. [1965b] On two systems of set theory. *Proc. Koningl. Nederl. Ak. v. Wetens. serie A* **68** pp. 95–9.

Da Costa, N.C.A. [1969] On a set theory suggested by Dedecker and Ehresmann I and II. *Proceedings of the Japan Academy* **45** pp. 880–8.

Da Costa, N.C.A. [1971] Remarques sur le système NF_1. *Comptes Rendus hebdomadaires des séances de l'Académie des Sciences de Paris série A* **272** pp. 1149–51.

Da Costa, N.C.A. [1974] Remarques sur les Calculs C_n, C_n^*, $C_n^=$, et D_n. *Comptes Rendus hebdomadaires des séances de l'Académie des Sciences de Paris série A* **278** pp. 818–21.

Dzierzgowski, D. [1992] Intuitionistic typical ambiguity. *Archive for Mathematical Logic* **31** pp. 171–82.

Dzierzgowski, D. [1993a] Typical ambiguity and elementary equivalence. *Mathematical Logic Quarterly* (formerly ZML) **39** pp. 436–446.

Dzierzgowski, D. [1993b] Le théorème d'ambiguïté et son extension à la logique intuitionniste. Dissertation doctorale. Université catholique de Louvain, Institut de mathématique pure et appliquée. Janvier 1993.

Dzierzgowski, D. [199?] Models of intuitionistic TT and NF. *Journal of Symbolic Logic* (to appear).

Engeler, E. and Röhrli, H. [1969] On the problem of foundations of category theory. *Dialectica* **23** pp. 58–66.

Feferman, S. [1972] Some formal systems for the unlimited theory of structures and categories (unpublished).

Forster, T.E. [1976] N.F. Ph.D. thesis, University of Cambridge.

* Forster, T.E. [1982] Axiomatising set theory with a universal set. In "La théorie des ensembles de Quine", *Cahiers du Centre de Logique*, Louvain-la-neuve **4** pp. 61–76.

* Forster, T.E. [1983a] *Quine's New Foundations, an introduction. Cahiers du Centre de Logique*, Louvain-la-neuve **5**.

Forster, T.E. [1983b] Further consistency and independence results in NF obtained by the permutation method. *Journal of Symbolic Logic* **48** pp. 236–8.

* Forster, T.E. [1985] The status of the axiom of choice in set theory with a universal set. *Journal of Symbolic Logic* **50** pp. 701–7. (The definition of "$\hat{<}\Phi$" is faulty in this paper.)

* Forster, T.E. [1987a] Permutation models in the sense of Rieger–Bernays. *Zeitschrift für mathematische Logik und Grundlagen der Mathematik* **33** pp. 201–10. (Theorem 2.3 is misstated: the correct version is theorem 3.1.30 here.)

* Forster, T.E. [1987b] Term models for weak set theories with a universal set. *Journal of Symbolic Logic* **52** pp. 374–87.

Forster, T.E. [1989] A second-order theory without a (second-order) model. *Zeitschrift für mathematische Logik und Grundlagen der Mathematik* **35** pp. 285–6.

Forster, T.E. [1992] On a problem of Dzierzgowski. *Bulletin de la Societé Mathématique de Belgique série B* **44** pp. 207–14.

Forster, T.E. [1993] A semantic characterisation of the well-typed formulae of λ-calculus. *Theoretical Computer Science* **110** pp. 405–8.

* Forster, T.E. and Kaye, R. [1991] End-extensions preserving power set. *Journal of Symbolic Logic* **56** pp. 323–8.[52]

* Forti, M. and Hinnion, R. [1989] The consistency problem for positive comprehension principles. *Journal of Symbolic Logic* **54** pp. 1401–18.

* Grishin, V.N. [1969] Consistency of a fragment of Quine's NF system. *Soviet Mathematics Doklady* **10** pp. 1387–90.

* Grishin, V.N. [1972a] The equivalence of Quine's NF system to one of its fragments (in Russian). *Nauchno-tekhnicheskaya Informatsiya* (series 2) **1** pp. 22–4.

* Grishin, V.N. [1972b] Concerning some fragments of Quine's NF system (in Russian). *Issledovania po matematicheskoy lingvistike, matematicheskoy logike i informatsionym jazykam* (Moscow) pp. 200–12.

* Grishin, V.N. [1972c] The method of stratification in set theory (in Russian). Ph.D. thesis, Moscow University.

* Grishin, V.N. [1973a] The method of stratification in set theory (abstract of thesis in Russian). Academy of Sciences of the USSR (Moscow) 9pp.

* Grishin, V.N. [1973b] An investigation of some versions of Quine's systems. *Nauchno-tekhnicheskaya Informatsiya* (series 2) **5** pp. 34–7.

* Hailperin, T. [1944] A set of axioms for logic. *Journal of Symbolic Logic* **9** pp. 1–19.

[52] Errata. p. 327. Line 11 should read 'and $a \in M$ such that $M \models \overline{\overline{\pi`a}} = \overline{\overline{\mathcal{P}`a}}$'. Line 13 the expression following '$M \models$' should be '$\overline{\overline{\pi`a}} = \overline{\overline{\mathcal{P}`a}}$'. Line 26 '(not just $\pi`a = \mathcal{P}`a$)' should read '(not just $\overline{\overline{\pi`a}} = \overline{\overline{\mathcal{P}`a}}$)'. Line 28 '$\pi`a$' should read '$\overline{\overline{\mathcal{P}`\pi`a}}$'.

Hatcher, W.S. [1963] La notion d'équivalence entre systèmes formels et une généralisation du système dit "New Foundations" de Quine. *Comptes Rendus hebdomadaires des séances de l'Académie des Sciences de Paris série A* **256** pp. 563–6.

* Henson, C.W. [1969] Finite sets in Quine's New Foundations. *Journal of Symbolic Logic* **34** pp. 589–96.

* Henson, C.W. [1973a] Type-raising operations in NF. *Journal of Symbolic Logic* **38** pp. 59–68.

* Henson, C.W. [1973b] Permutation methods applied to NF. *Journal of Symbolic Logic* **38** pp. 69–76.

* Hiller, A.P. and Zimbarg, J.P. [1984] Self-reference with negative types. *Journal of Symbolic Logic* **49** pp. 754–73.

Hinnion, R. [1972] Sur les modèles de NF. *Comptes Rendus hebdomadaires des séances de l'Académie des Sciences de Paris série A* **275** p. 567.

* Hinnion, R. [1974] Trois résultats concernant les ensembles fortement cantoriens dans les "New Foundations" de Quine. *Comptes Rendus hebdomadaires des séances de l'Académie des Sciences de Paris série A* **279** pp. 41–4.

* Hinnion, R. [1975] Sur la théorie des ensembles de Quine. Ph.D. thesis, ULB Brussels.

* Hinnion, R. [1976] Modèles de fragments de la théorie des ensembles de Zermelo–Fraenkel dans les "New Foundations" de Quine. *Comptes Rendus hebdomadaires des séances de l'Académie des Sciences de Paris série A* **282** pp. 1–3.

* Hinnion, R. [1979] Modèle constructible de la théorie des ensembles de Zermelo dans la théorie des types. *Bulletin de la Societé Mathématique de Belgique série B* **31** pp. 3–11.

* Hinnion, R. [1980] Contraction de structures et application à *NFU*: Définition du "degré de non-extensionalité" d'une relation quelconque. *Comptes Rendus hebdomadaires des séances de l'Académie des Sciences de Paris série A* **290** pp. 677–80.

* Hinnion, R. [1981] Extensional quotients of structures and applications to the study of the axiom of extensionality. *Bulletin de la Societé Mathématique de Belgique série B* **33** pp. 173–206.

* Hinnion, R. [1982] NF et l'axiome d'universalité. In "La théorie des ensembles de Quine", *Cahiers du Centre de Logique*, Louvain-la-neuve **4** pp. 45–59.

* Hinnion, R. [1986] Extensionality in Zermelo–Fraenkel set theory. *Zeitschrift für mathematische Logik und Grundlagen der Mathematik* **32** pp. 51–60.

* Hinnion, R. [1989] Embedding properties and anti-foundation in set theory. *Zeitschrift für mathematische Logik und Grundlagen der Mathematik* **35** pp. 63–70.

* Hinnion, R. [1990] Stratified and positive comprehension seen as superclass rules over ordinary set theory. *Zeitschrift für mathematische Logik und Grundlagen der Mathematik* **36** pp. 519–34.

* Holmes, M.R. [1991a] Systems of combinatory logic related to Quine's 'New Foundations'. *Annals of Pure and Applied Logic* **53** pp. 103–33.

* Holmes, M.R. [1991b] The Axiom of Anti-Foundation in Jensen's 'New Foundations with Ur-Elements' *Bulletin de la Société Mathématique de Belgique, série B* **43** no. 2, pp. 167–79.

* Holmes, M.R. [1992] Modelling fragments of Quine's 'New Foundations', *Cahiers du Centre de Logique* **7**, Louvain-la-Neuve, pp. 97-112.

* Holmes, M.R. [1993] Systems of combinatory logic related to predicative and 'mildly impredicative' fragments of Quine's 'New Foundations'.*Annals of Pure and Applied Logic* **59** pp. 45–53.

* Holmes, M.R. [1994] The set theoretical program of Quine succeeded (but nobody noticed). *Modern Logic* **4** pp. 1–47.

* Jamieson, M.W. [1994] Set theory with a Universal Set. Ph.D. thesis, University of Florida.

* Jensen, R.B. [1969] On the consistency of a slight(?) modification of Quine's NF. *Synthese* **19** pp. 250–63.

* Kaye, R.W. [1991] A generalisation of Specker's theorem on typical ambiguity. *Journal of Symbolic Logic* **56** pp. 458–66.

Kaye, R.W. [199?] The quantifier complexity of *NF*. (to appear.)

Kemeny, J. G. [1950] Type theory *vs* set theory (abstract). *Journal of Symbolic Logic* **15** p. 78.

Kirmayer, G. [1981] A refinement of Cantor's theorem. *Proceedings of the American Mathematical Society* **83** p. 774.

* Körner, F. [1994] Cofinal indiscernibles and some applications to New Foundations. *Mathematical Logic quarterly* **40** pp. 347–56.

Kühnrich, M. and Schultz, K. [1980] A hierarchy of models for Skala's set theory. *Zeitschrift für mathematische Logik und Grundlagen der Mathematik* **26** pp. 555–9.

Kuzichev, A.C. [1981] Arithmetic theories constructed on the basis of λ-conversion. *Soviet Mathematics Doklady* **24** pp. 584–9.

Kuzichev, A.C. [1983] Nyeprotivoretchivost' Sistema NF Quine. *Doklady Akademia Nauk* **270** pp. 537–41.

Lake, J. [1974] Some topics in set theory. Ph.D. thesis, Bedford College, London University.

Lake, J. [1975] Comparing type theory and set theory. *Zeitschrift für mathematische Logik und Grundlagen der Mathematik* **21** pp. 355–6.

McLarty, C. [1992] Failure of cartesian closedness in *NF*. *Journal of Symbolic Logic* **57** pp. 555–6.

McNaughton, R. [1953] Some formal relative consistency proofs. *Journal of Symbolic Logic* **18** pp. 136–44.

* Malitz, R.J. [1976] Set theory in which the axiom of foundation fails. Ph.D. thesis, UCLA.

Manakos, J. [1984] On Skala's set theory. *Zeitschrift für mathematische Logik und Grundlagen der Mathematik* **30** pp. 541–6.

* Mitchell, E. [1976] A model of set theory with a universal set. Ph.D. thesis, University of Wisconsin, Madison, Wisconsin.

Oberschelp, A. [1964] Eigentliche Klasse als Urelemente in der Mengenlehre. *Mathematische Annalen* **157** pp. 234–60.

Oberschelp, A. [1973] Set theory over classes. *Dissertationes Mathematicae* **106** 62pp.

Orey, S. [1955] Formal development of ordinal number theory. *Journal of Symbolic Logic* **20** pp. 95–104.

Orey, S. [1956] On the relative consistency of set theory. *Journal of Symbolic Logic* **21** pp. 280–90.

* Orey, S. [1964] New Foundations and the axiom of counting . *Duke Mathematical Journal* **31** pp. 655–60.

* Oswald, U. [1976] Fragmente von "New Foundations" und Typentheorie. Ph.D. thesis, ETH Zürich.

* Oswald, U. [1981] Inequivalence of the fragments of New Foundations. *Archiv für mathematische Logik und Grundlagenforschung* **21** pp. 77–82.

* Oswald, U. [1982] A decision method for the existential theorems of NF_2. *Cahiers du Centre de Logique* Louvain-la-neuve **4** pp. 23–43.

* Oswald, U. and Kreinovich, V. [1982] A decision method for the Universal sentences of Quine's NF. *Zeitschrift für mathematische Logik und Grundlagen der Mathematik* **28** pp. 181–7.

* Pabion, J.F. [1980] TT_3I est équivalent à l'arithmétique du second ordre. *Comptes Rendus hebdomadaires des séances de l'Académie des Sciences de Paris série A* **290** pp. 1117–18.

* Pétry, A. [1974] À propos des individus dans les "New Foundations" de Quine. *Comptes Rendus hebdomadaires des séances de l'Académie des Sciences de Paris série A* **279** pp. 623–4.

* Pétry, A. [1975] Sur l'incomparabilité de certains cardinaux dans le "New Foundations" de Quine. *Comptes Rendus hebdomadaires des séances de l'Académie des Sciences de Paris série A* **281** pp. 673–5.

* Pétry, A. [1976] Sur les cardinaux dans le "New Foundations" de Quine. Ph.D. thesis, University of Liège.

* Pétry, A. [1977] On cardinal numbers in Quine's NF. *Set theory and hierarchy theory.* Springer Lecture Notes in Mathematics **619** pp. 241–50.

* Pétry, A. [1979] On the typed properties in Quine's NF. *Zeitschrift für mathematische Logik und Grundlagen der Mathematik* **25** pp. 99–102.

* Pétry, A. [1982] Une charactérisation algébrique des structures satisfaisant les mêmes sentences stratifiées. *Cahiers du Centre de Logique* Louvain-La-Neuve **4** pp. 7–16.

* Pétry, A. [1992] Stratified languages. *Journal of Symbolic Logic* **57** pp. 1366–76.

Prati, N. [1994] A partial model of *NF* with E. *Journal of Symbolic Logic* **59** pp. 1245–53.

Quine, W.v.O. [1937a] New foundations for mathematical logic. *American Mathematical Monthly* **44** pp. 70–80 (reprinted in Quine [1953a]).

Quine, W.v.O. [1937b] On Cantor's theorem. *Journal of Symbolic Logic* **2** pp. 120–4.

Quine, W.v.O. [1945] On ordered pairs. *Journal of Symbolic Logic* **10** pp. 95–6.

Quine, W.v.O. [1951a] *Mathematical logic* (2nd ed.) Harvard University Press.

Quine, W.v.O. [1951b] On the consistency of "New Foundations". *Proceedings of the National Academy of Sciences of the USA* **37** pp. 538–40.

Quine, W.v.O. [1953a] *From a logical point of view.* Harper & Row, London.

* Quine, W.v.O. [1953b] On ω-inconsistency and a so-called axiom of infinity. *Journal of Symbolic Logic* **18** pp. 119–24. Reprinted in Quine, [1966].

* Quine, W.v.O. [1966] *Selected logic papers.* Random House, New York.

* Quine, W.v.O. [1967] *Set theory and its logic.* Belknap Press.

Quine, W.v.O. [1987] The inception of *NF*. *Bulletin de la Société Mathématique de Belgique* série B **45** pp. 325–28.

Rosser, J. B. [1939a] On the consistency of Quine's new foundations for mathematical logic. *Journal of Symbolic Logic* **4** pp. 15–24.

Rosser, J. B. [1939b] Definition by induction in Quine's New Foundations for Mathematical Logic. *Journal of Symbolic Logic* **4** p. 80.

Rosser, J. B. [1942] The Burali-Forti paradox. *Journal of Symbolic Logic* **7** pp. 11–17.

Rosser, J. B. [1952] The axiom of infinity in Quine's New Foundations. *Journal of Symbolic Logic* **17** pp. 238–42.

Rosser, J. B. [1953a] *Logic for mathematicians.* McGraw-Hill, reprinted (with appendices) by Chelsea, New York, 1978.

Rosser, J. B. [1953b] *Deux esquisses de logique.* Paris.

* Rosser, J. B. [1954] Review of Specker [1953]. *Journal of Symbolic Logic* **19** p. 127.

Rosser, J. B. [1956] The relative strength of Zermelo's set theory and Quine's new foundations. *Proceedings of the International Congress of Mathematicians (Amsterdam 1954)* III. pp. 289–94.

* Rosser, J. B. [1978] second edition of [1953a]. Chelsea, New York.

* Rosser, J. B. and Wang, H. [1950] Non-standard models for formal logic. *Journal of Symbolic Logic* **15** pp. 113–29.

* Russell, B. A. W. [1908] Mathematical logic as based on the theory of types. *American Journal of Mathematics* **30** pp. 222–62.

* Russell, B. A. W and Whitehead, A. N. [1910] *Principia mathematica.* Cambridge University Press.

Schultz, K. [1977] Ein Standardmodell für Skala's Mengenlehre. *Zeitschrift für mathematische Logik und Grundlagen der Mathematik* **23** pp. 405–8.

Schultz, K. [1980] The consistency of *NF*. Unpublished typescript: copies in the possession of members of the Séminaire NF.

* Scott, D. S. [1960] Review of Specker [1958]. *Mathematical Reviews* **21** p. 1026.

* Scott, D. S. [1962] Quine's individuals. In *Logic, methodology and philosophy of science.* Ed. E. Nagel, pp. 111–5, Stanford University Press.

* Scott, D. S. [1980] The lambda calculus: some models, some philosophy. *The Kleene Symposium.* North-Holland, Amsterdam, pp. 116–24.

Sheridan, K. J. [199?] The singleton function is a set in a slight extension of Church's set theory. D.Phil. thesis, University of Oxford.

Skala, H. [1974a] Eine neue Methode, die Paradoxien der naiven Mengenlehre zu vermeiden. *Annalen der Österreichen Akademie der Wissenschaften Math-Nat. Kl. II.* pp. 15–16.

Skala, H. [1974b] An alternative way of avoiding the set-theoretical paradoxes. *Zeitschrift für mathematische Logik und Grundlagen der Mathematik* **20** pp. 233–7.

* Specker, E. P. [1953] The axiom of choice in Quine's new foundations for mathematical logic. *Proceedings of the National Academy of Sciences of the USA* **39** pp. 972–5.

* Specker, E. P. [1958] Dualität. *Dialectica* **12** pp. 451–65.

* Specker, E. P. [1962] Typical ambiguity. In *Logic, methodology and philosophy of science.* Ed E. Nagel, Stanford University Press.

Stanley, R. L. [1955] Simplified foundations for mathematical logic. *Journal of Symbolic Logic* **20** pp. 123–39.

* Wang, H. [1950] A formal system of logic. *Journal of Symbolic Logic* **15** pp. 25–32.

Wang, H. [1952a] Negative types. *MIND* **61** pp. 366–8.

* Wang, H. [1953] The categoricity question of certain grand logics. *Mathematische Zeitschrift* **59** pp. 47–56.

* Weydert, E. [1989] How to approximate the naïve comprehension scheme inside of (*sic*) classical logic. Ph.D. Thesis, Friedrich-Wilhelms-Universität Bonn. Published as *Bonner mathematische Schriften* **194**.

Concerning other matters raised in the text

Bernays, P. [1954] A system of axiomatic set theory VII. *Journal of Symbolic Logic* **19** pp. 81–96.

Boolos, G. [1979] *The unprovability of consistency.* Cambridge University Press.

Chang, C.C. and Keisler, H.J. [1973] *Model Theory.* North-Holland, Amsterdam.

Church, A. [1940] A formulation of the simple theory of types. *Journal of Symbolic Logic* **5** pp. 56–68.

Degen, J.W. [1988] There can be a permutation which is not the product of two reflections. *Zeitschrift für mathematische Logik und Grundlagen der Mathematik* **34** pp. 65–6.

Fine, K. [1978] Model theory for modal logic II: the elimination of *de re* modality. *Journal of Philosophical Logic* **7** pp. 277–306.

Forti, M. [1987] Models of the generalized positive comprehension principle. Preprint, Università di Pisa.

Forti, M. and Honsell, F. [1983] Set theory with free construction principles. *Annali della Scuola Normale Superiore di Pisa, Scienze fisiche e matematiche* **10** pp. 493–522.

Forti, M. and Honsell, F. [1984a] Axioms of choice and free construction principles I. *Bulletin de la Société Mathématique de Belgique série B* **36** pp. 69–79.

Forti, M. and Honsell, F. [1984b] Comparison of the axioms of local and global universality. *Zeitschrift für mathematische Logik und Grundlagen der Mathematik* **30** pp. 193–6.

Gaughan, E. [1967] Topological group structures of infinite symmetric groups. *Proceedings of the National Academy of Sciences of the USA* **53** pp. 907–10.

Gödel, K. [1940] *The consistency of the continuum hypothesis.* Princeton University Press.

Kripke, S. [1975] Outline of a theory of truth. *Journal of Philosophy* **72** pp. 690–716.

Levy, A. [1965] A hierarchy of formulae in set theory. *Memoirs of the American Mathematical Society* no. 57.

McDermott, M. [1977] Sets as open sentences. *American Philosophical Quarterly* **14** pp. 247–53.

Macpherson, H. D. and Neumann, P. [1990] Subgroups of infinite symmetric groups. *Journal of the London Mathematical Society* **42** pp. 64–84.

Mathias, A.R.D. [199?] Notes on MacLane set theory. *unpublished.* To appear.

Powell, W.J. [1975] Extending Gödel's negative interpretation to *ZF*. *Journal of Symbolic Logic* **40** pp. 221–9.

Quine, W.v.O. [1986] Peano as Logician. Celebrazioni in memoria di Giuseppe Peano nel cinquantenario della morte. Atti del Convegno organizzato dal Dipartimento de Matematica dell'Università di Torino 27–28 ottobre 1982. Torino 1986.

Rieger, L. [1957] A contribution to Gödel's axiomatic set theory. *Czechoslovak Mathematical Journal* **7** pp. 323–57.

Scroggs, S.J. [1951] Extensions of the Lewis system S5. *Journal of Symbolic Logic* **16** pp. 112–20.

Vopénka, P. [1979] *Mathematics in alternative set theory.* Teubner-Texte, Leipzig.

Vopénka, P. and Hájek, P. [1972] *The theory of semisets.* North-Holland, Amsterdam.

Wang, H. [1952b] Truth definitions and consistency proofs. *Transactions of the American Mathematical Society* **72** pp. 243–75.

INDEX OF DEFINITIONS

$\prec$ 7
$\Box$ 109
$\Diamond$ 109–11, 113–14, 143, 144, 146
$\hookrightarrow_e^{\mathcal{P}}$ 6
$\subseteq_e^{\mathcal{P}}$ 6
$\#$ 7
$<^T$ 32, 114
2^x 29
$\forall^*$, $\forall^n$, $\exists^*$, $\forall_2$, $\forall^*\exists^*$ 5

AC_2 42, 101, 102, 145, 146
AC_ω 101
AC_{wo} 52, 144
$Amb(\Gamma)$ 58
ambiguity 58
A-object 139
AxCount 30–3, 52–4, 97, 100, 102–4, 119, 155
AxCount$_\leq$ 31–3, 44, 52–4, 90, 107, 113, 119, 146
AxCount$_\geq$ 31

B, b 9
beth numbers 5, 7, 32, 49, 50, 52, 53, 64, 67, 90, 114
bisimulation 19
Boffa atom 9, 98, 101

canonical stratification 7
cantorian, $can(x)$ 27, 28, 31, 57, 58, 63, 119
$card$ 5
centralizer 88
complement object 139
constructible set 43, 55, 75, 82, 91, 145
contraction 19, 20, 72
CUS 11, 22, 23, 122, 136
cycle 5, 37, 38, 102, 103

Δ_0^{Levy} hierarchy 5, 44, 61, 75, 82, 104, 105, 146

$\in$-determinacy 13

f^T 51
fix 42

GC 101–3, 116

G_x 13
$G_{x=y}$ 19

H 109, 111
H_ϕ 7, 9, 34, 35, 56, 127
H_X 9
Hartogs' aleph function 47–8, 54, 83, 106
Hartogs' theorem 45, 48, 55
Hinnion's + operation 19
homogeneous 28–9, 34, 36–8, 40, 48, 50, 55–7, 62, 90, 100, 105, 112, 145

I, II 13
invariant 32, 97, 103–4, 109–13, 116–19
ι 9

j, J_n 8, 33, 37, 88, 101, 108, 111, 115

L_1‘X, L_2‘X 45
Λ 4

low comprehension 124, 125, 131–3, 137, 138, 141, 142

M^* 7
ML 46
m_x 7

$\mathbb{N}$ 28
NC 4
NCI 4, 52
n-equivalence 8, 33
NF v, 1–4, 8–13, 15, 20, 22–5, 27–31, 33–44, 46, 47, 50–6, 58, 60–7, 70, 73–5, 82–4, 87–92, 96–102, 104–9, 111, 115, 117, 119, 120, 142, 144–6, 148, 149, 151, 154–7
$NF\forall^*$ 145
NFC 11, 31, 36, 63, 104, 144, 146
NFU 10, 12, 25, 67–70, 72, 89, 141, 149, 153
NF_k, NF_2, NF_3 10, 25, 26, 58, 65, 83, 89, 105, 122, 123, 141, 149
$NF\Gamma$ 10

NFO 9, 10, 65, 83–5, 87–9, 130, 132, 133, 135, 136, 143
NO 4, 24, 44, 46–7, 51–2, 56, 74, 90, 120
No 44
n-symmetric 8

Ω 46

$\mathcal{P}$ 5
$\mathcal{P}$ hierarchy 5
power object 139
pseudoinduction 13, 18, 20, 39

Quine atom 9, 12, 20, 38, 84, 92, 99, 100, 106–9, 111, 127
Quine ordered pair 28, 97, 105, 109

replacement 7, 8, 18, 23, 24, 42, 93, 121, 139
ρ 48
$RUSC(R)$ 9

saturated boolean algebra 60
setlike 8, 33, 75, 93–7, 102, 105, 111, 115, 119, 121
Σ_X 9
Specker's Φ function 50–1, 54–5
stratified 7
  n-stratified 7
  weakly stratified 8
stratimorphism 93–5
strong extensionality (axiom of) 20, 104
strongly cantorian $stcan(x)$ 27, 28, 31, 57, 82, 91, 100, 101, 103
$str(T)$ 10

T 30
τ_n 96
$TC(x)$ 5
$\Theta`f$ 51
TNT 6, 53, 60, 61, 65, 72, 146, 147
TNTU 68
tsau 58
TST 6, 7, 22, 58–62, 65, 66, 68, 85, 86
TST_k 6, 62, 63
TSTI 7, 64, 73
$TSTI_k$ 62, 63
TSTU 68
type 7
typed 8, 106, 120

$V \in V$ v

wff 7
$WF(x)$ 40

Z 11
ZF 1–4, 7, 11, 22–5, 30, 40, 42, 43, 46, 54–6, 73, 82, 83, 89, 90, 92, 102, 106, 119, 121, 125, 126, 139, 141, 142, 144, 145
ZFC 11, 25, 55, 144

AUTHOR INDEX

Ackermann 113, 122
Aczel 2, 13

Barcan 109
Beneš 26
Bernays vi, 25, 92
Boffa 7, 25, 34, 40, 56, 60, 66, 68, 82, 83, 89, 106, 111, 141
Boolos 109

Casalegno 89
Church vi, 2, 7, 9, 11, 22–5, 126
Coret 7
Crabbé 25, 27, 49, 60, 64, 89

Degen 102
Dzierzgowski 25, 60, 128, 142, 151

Feferman 72
Fine 109, 110
Forster 12, 25, 49, 50, 56, 64, 72, 87, 89, 101, 112, 118
Forti v, 9, 11, 13, 19–21

Gaughan 108
Gödel 4, 89
Grishin 25, 60, 65, 141

Hailperin 25, 26, 63, 72
Hájek 31
Henson 25, 30, 33, 49–51, 58, 90, 93, 96, 98, 99, 101, 105, 109, 119
Hinnion v, 19–21, 25, 45, 72, 83, 84, 89, 90, 98, 100, 145
Honsell 9, 11, 20
Hyland 83

Jensen 25, 141

Kaye v, vi, 60, 65, 72, 73, 83
Kemeny 73
Körner 119
Kripke 12
Kuzichev 26

Lake 27, 72
Levy 5, 6, 25

Maciocia 108
Macpherson 88
Malitz 23, 25
Mathias 7, 72, 98
McDermott 12
Mitchell 3, 9, 11, 23–5, 139

Neumann 88

Orey 10, 25, 33, 62, 66, 89
Oswald 10, 25, 89, 122

Pétry 25, 32, 33, 37, 38, 93, 94, 98, 100, 106, 112, 119, 120

Quine 3, 4, 9, 11–13, 22, 25, 28, 29, 46, 73, 107

Rieger vi, 25, 92
Rosser 4, 9, 29–31, 46, 47, 74, 89, 91, 111, 146
Russell 1, 2, 4, 6, 11, 12, 22–4, 27, 51

Schultz, 26
Scott 25, 92, 97
Scroggs 109
Sheridan vi, 23, 141
Skala v, 3, 25
Specker 22, 25, 49–52, 54, 61, 68, 90

Vopénka 31

Wang 6 38, 46, 47, 62, 74
Weydert v, 3, 21, 23

Yasuhara 101

GENERAL INDEX

abstract, set and class 1, 8, 9, 33–9, 46, 50, 55, 66, 71, 115
AC_2 42, 101, 102, 145, 146
AC_ω 101
AC_{wo} 52, 144
$[\alpha]$ 48
antimorphism 12, 98
automorphism 8, 30, 31, 33, 34, 36, 41, 42, 58, 59, 68–70, 82, 83, 92, 93, 97, 101–3, 107, 119, 121, 143, 145, 146
axiom of choice 11, 23, 25, 36, 44, 50, 51, 53, 57, 61, 67, 75, 82, 83, 90, 101, 102, 116, 137, 145, 146
$\mathrm{AxCount}_{\leq}$ 31–3, 44, 52–4, 90, 107, 113, 119, 146
axiom of counting 25, 30–3, 43, 52–4, 97, 100, 102–4, 119, 155
axiom of foundation 4, 12, 18, 23, 36, 42, 73, 75, 92, 139, 149, 155
axiom of infinity 7, 23, 28–31, 44, 49, 53, 64, 65, 67, 84, 86, 114, 156, 157
axiom of power set 75, 145
axiom of sumset 72, 73, 75, 79, 145

base set 92, 93, 97, 117
beth numbers 5, 7, 32, 49, 50, 52, 53, 64, 67, 90, 114
BF 90, 100, 101
bisimulation 19
Boffa atom 9, 98, 101
Boffa's W 56, 147
boolean algebra 9, 10, 26, 60, 61, 85, 86
Bounding Lemma 73, 74
Burali-Forti paradox 24, 44, 46, 48, 74, 100, 144

cantorian 27, 28, 31, 57, 58, 63, 119
Cantor's paradox 2, 24, 74, 144
Cantor's theorem 24, 27, 32, 37, 38, 154, 156
centralizer 88
Church's set theory (*CUS*) 11, 22, 23, 122, 136

comprehension 1, 3, 6, 10, 22, 24, 26, 29, 30, 55, 57, 65, 68–70, 89, 125, 127, 137, 141, 145
consistency proof 3, 25–7, 62, 64, 75, 89, 122, 127, 142, 155
consistency strength 70, 90, 144
constructible set 43, 55, 75, 82, 91, 145
contraction 19, 20, 72
cycle 5, 37, 38, 102, 103

$\Delta_n^{\mathcal{P}}$ formulae 5, 8, 43, 44, 72, 73, 142
direct limit 85, 86, 105
duality scheme 12, 107, 143

Ehrenfeucht–Mostowski theorem vi, 70, 82, 119–21
embedding 84, 85, 93, 115, 117, 118
  $\mathcal{P}$-embedding 6, 105, 126–8
end-extension 5, 6, 9, 46, 61, 104–6
existence property 83, 84
extensionality 6, 10, 12, 18, 19, 24, 26, 67, 72, 75, 84, 87, 91, 93, 94, 104, 123, 125, 128, 130, 133, 135, 137, 139

finitely generated model 85–7
finitism 52

GC 101–3, 116
Gödel number 38, 62, 63

H_ϕ 7, 9, 34, 35, 56, 127
Hailperin tuples 75
Hartogs' aleph function 47, 48, 54, 83, 106
Hartogs' theorem 45, 48, 55
hereditarily finite set 34, 39, 54, 123
hereditarily low 127, 129, 136
homogeneous 28, 29, 34, 36–8, 40, 48, 50, 55–7, 62, 90, 100, 105, 112, 145

inductive definitions 57, 141
invariant 32, 97, 103, 104, 109–13, 116–19
involution 102, 107, 116

J_n 8, 33, 37, 88, 101, 108, 111, 115

*K*5 109, 111
Kaye's Lemma 60, 65, 72
KF 43, 72, 73, 75, 82, 83, 145

λ-calculus 1, 2, 7, 29, 154
large cardinal 4, 144
Levy hierarchy 5, 44, 61, 75, 82, 104, 105, 146
Logicism 1, 11, 12, 18, 19
low comprehension 124, 125, 131–3, 137, 138, 141, 142
L-strings 91

many-sorted 93, 94
Mirimanoff's paradox 24, 144
modal logic 92, 93, 109–11, 119
Mostowski's collapse lemma 45, 90, 100

natural deduction 8
NC 4, 30, 44, 54, 90, 120
NCI 4, 52
n-equivalent 8, 33
NF v, 1–4, 8–13, 15, 20, 22–5, 27–31, 33–44, 46, 47, 50–6, 58, 60–7, 70, 73–5, 82–4, 87–92, 96–102, 104–9, 111, 115, 117, 119, 120, 142, 144–6, 148, 149, 151, 154–7
NF_2 10, 83, 122, 123
NF_3 10, 25, 26, 58, 65, 89, 105, 141, 149
NF_4 9, 65
NFC 11, 31, 36, 63, 104, 144, 146
NFO 9, 10, 65, 83–5, 87–9, 130, 132, 133, 135, 136, 143
n-formula 8, 33, 34
NFU 10, 12, 25, 67–70, 72, 89, 141, 149, 153
NO 4, 24, 44, 46, 47, 49, 51–3, 56, 74, 90, 120
$\exists NO$ 74, 75, 83, 144
n-stratified formula 7, 63, 64, 97

Ω 47, 56
Orey model 64

$\mathcal{P}$-extension 6, 126, 130, 132, 136, 141
permutation model 82, 92–5, 97–101, 103–9, 111, 114, 115, 117, 118, 121, 126, 127, 129, 143, 145, 146
$\Pi^{\mathcal{P}}_n$ formulae 5, 43, 44, 61, 74, 105
positive set theory v, 3, 24
predication 1, 11, 12
preservation theorem 8, 93, 95, 96
pseudofoundation 34, 140, 142
pseudoinduction 13, 18, 20, 39

Quine atom 9, 12, 20, 38, 84, 92, 99, 100, 106–9, 111, 127
Quine ordered pairs 28, 97, 105, 109

rank 17, 20, 34, 48, 53, 54, 68–70, 72, 85, 87, 88, 127, 135, 137, 144
relational type 30, 90, 100, 101, 112
replacement 7, 8, 18, 23, 24, 42, 93, 121, 139
rudimentary function 81, 82
Russell's paradox 1–3, 22, 24, 27, 144

S4 109
self-membered 12, 84, 88, 98, 107, 108, 127, 143
separation 24, 47, 69, 70, 72–4, 76, 82, 125, 141, 145
setlike permutation 8, 33, 75, 93–7, 102, 105, 111, 115, 119, 121
$\Sigma^{\mathcal{P}}_n$-formulae 5, 6, 44, 61, 72, 114, 146
skew-conjugation 108, 115, 116
small 21, 57, 58, 88, 89, 143
Specker's ϕ function 50–1, 54–5
standard integer 29–31, 33, 34, 42, 43, 47, 50, 51, 53, 54, 58, 64, 68, 74, 115, 146
stratimorphism 93–5
strong extensionality 20, 104
strongly cantorian 27, 28, 31, 57, 82, 91, 100, 101, 103
support 108, 111
symmetric group 8, 9, 72, 88, 93, 102, 109, 115
symmetric set 8, 39, 88, 89, 107, 112, 117, 143, 146

$<^T$ 32, 114
term model 10, 15, 34, 38, 39, 83, 84, 87, 88, 107, 115, 122, 136, 143, 145, 146
TNT 6, 53, 60, 61, 65, 72, 146, 147
TNTU 68
topology 3, 12, 17, 19, 21, 22, 24, 108, 143
transitive set 5, 20, 34–6, 39, 40, 42, 73–5, 83, 90, 100, 104, 105, 126, 145, 146
transposition 5, 37, 38, 92, 99–101, 104–7, 110
tree 30, 48, 49, 53, 54
truth definition 8, 12, 25, 61, 62, 66

TST 6, 7, 22, 58–62, 65, 66, 68, 85, 86
TSTI 7, 64, 73
$TSTI_k$ 62, 63
TST_k 6, 62, 63
TSTU 68
typed formula 8, 106, 120

ultrafilter 35, 39
ultrafinitism 52
ultraproduct 61

V_ω vii, 40, 145

weakly stratified 8, 98, 112
well-founded extensional relation 33, 35, 37, 45, 83, 90, 91, 100, 145
well-founded relation 12, 17, 32, 48, 52, 88–90, 101, 114, 126
well-founded set 2–4, 17, 23, 24, 27, 34, 36, 40–4, 55, 73, 74, 83, 107, 113–15, 121, 125–8, 130, 138, 141, 142
well-quasi-order 32, 36
well-order 5, 38, 44, 45, 47, 48, 51–3, 55–8, 74, 82, 85, 91, 120, 144, 145
Wiener–Kuratowski 26, 27, 29, 75, 80, 81, 97, 109, 136
WO 56

Zermelo set theory 7, 11, 12, 52, 54, 68–70, 73, 75, 90, 132, 141
ZF 1–4, 7, 11, 22–5, 30, 40, 42, 43, 46, 54–6, 73, 82, 83, 89, 90, 92, 102, 106, 119, 121, 125, 126, 139, 141, 142, 144, 145
ZFC 11, 25, 55, 144